ARABIAN PLATE HYDROCARBON GEOLOGY AND POTENTIAL— A Plate Tectonic Approach

by Z. R. Beydoun

American University of Beirut, Beirut, Lebanon
and
Marathon International Petroleum (GB) Ltd., London, U.K.

AAPG Studies in Geology #33

Published by
The American Association of Petroleum Geologists
Tulsa, Oklahoma, U.S.A.

Published July 1991

ISBN: 0-89181-041-2
ISSN: 0149-1377

Association Editor: Susan A. Longacre
Science Director: Gary D. Howell
Publications Manager: Cathleen P. Williams
Special Projects Editor: Anne H. Thomas

Cover photos by Gary D. Howell.
Cover design by Elizabeth A. Mitchell.

*To my wife Muntaha Saghieh, Levant/Mesopotamian archaeologist,
in recognition of her professional dedication, the example she sets,
and the encouragement and support she provides.*

About the Author

Ziad Rafik Beydoun was born in Beirut, Lebanon, in December 1924, and was brought up in Palestine where he had his primary and secondary schooling. His geological education was undertaken at the University of Oxford, from which he received his B.A. in Geology in 1948 (Honour School of Natural Science) and qualified for the M.A. in 1953.

On obtaining his B.A. degree, Dr. Beydoun joined the Iraq Petroleum Company (IPC) and Associated Companies Group, in September 1948. He worked throughout their areas of activity extending from Lebanon and the Levant through Syria and Iraq to the Gulf (Qatar, Bahrain, the Trucial States—now the United Arab Emirates), Oman and the Aden Protectorate (later South Yemen and now part of the Republic of Yemen). The first five years were divided between surface mapping and subsurface geology, but from 1953 his main activity concentrated in surface work in the Aden Protectorate. Between 1959 and 1961, he had several periods of absence to work on his doctoral dissertation (Aden Protectorate geology) at Oxford. He was awarded the Doctor of Philosophy degree in June 1961; this was subsequently published in 1964 by the British Overseas Geological Survey as a Bulletin Supplement.

In the summer of 1961, he was seconded to Partex (the Gulbenkian interests in Oman and the Gulf) as Representative based in Qatar, and he participated in the discovery of oil in Oman by PDO.

In 1963 he left Partex and the service of the IPC group and returned to Lebanon for family reasons. There he joined the Ministry of National Economy as "Expert" on mineral and petroleum resources and was actively engaged in assessing all exploration licenses in the country. Shortly afterward, he also became Assistant Professor and later Associate Professor of Geology and Chairman of the Department of Geology, American University of Beirut (AUB).

Dr. Beydoun's association with Marathon Oil Company dates to 1964, when he consulted for the newly opened exploration office in London in the summers of 1964 and 1965. He joined Marathon International Petroleum (GB) Ltd. in London on a full-time basis in 1966 as a Senior (review) Geologist responsible for the Middle East. In the summer of 1970, he returned to Lebanon for personal reasons and was appointed Professor of Geology and Chairman of the Department at the American University of Beirut. He also continued as Middle East Geological Adviser for Marathon, spending consecutive summers in the London office and occasionally at the company's Denver Research Center, as well as being actively involved during the academic year with the company's Middle East activities, especially in Syria.

In July 1985 he took two years' leave of absence from the American University of Beirut to return to full-time duties with Marathon in London. On expiry of this, he was seconded by the University to the UNDP/World Bank Red Sea-Gulf of Aden Regional Hydrocarbon Study Project as the Project Scientific Advisor, operating from his London base with Marathon on behalf of AUB. He is currently New Ventures Representative/Manager and Middle East Geological Adviser with Marathon in London, and Adjunct Professor of Geology at the American University of Beirut.

Dr. Beydoun continues to be actively involved in Middle East geology, with frequent professional visits to different countries of the region. He has over 25 scientific publications on the geology of portions of or on the Middle East region, including the well-reviewed "The Middle East: Regional Geology and Petroleum Resources" published in the U.K. in 1988 and now available on CD-ROM through the Masera-AAPG Data Systems. Many of his former graduates now hold key positions in the oil companies and earth resource ministries of the Middle East, and this constitutes a great source of pride and joy to him as well as a strong bond of friendships criss-crossing the region.

Contents

Foreword

This study developed from a keynote paper that had been invited for the "Geodynamics of the Arabian Plate," an international conference organized by the Department of Geology, University of Kuwait, and the Kuwait Foundation for the Advancement of Sciences, which had been scheduled for October 1990 in Kuwait but, regrettably, could not take place. The original paper was entitled "The hydrocarbon potential of the Arabian plate: its relation to plate margin history and intra-plate structuring." Only an extended abstract was submitted, but a draft of the essentials of the theme was in progress.

Since the topic is especially timely, colleagues have urged me to develop the theme and expand it into a long essay review, in view of my long and varied first-hand experience in the region over four decades. I have found the idea challenging and I hope that I have achieved what I had intended by providing in the following pages a scientifically accurate overview of what lies behind the region's special richness in hydrocarbons. In putting forward some novel suggestions that will help focus on new exploration plays and thus unlock additional reserves, my additional hope is that these might also serve as analogs elsewhere.

January 1991

Abstract

Reported proven hydrocarbon reserves of the Arabian plate region at the start of 1991 totaled 663.2 billion barrels (B bbl) of oil and 1,325.4 trillion cubic feet (tcf) of gas (66.4% and 31.5% of the world's oil and gas reserves, respectively). More than 98% of these are concentrated in the northeast margin region between northwest Iraq and Central Oman and lie in reservoirs ranging in age from late Paleozoic to early Neogene. Additional reserves, however, increasingly are being established along the other Arabian plate margins and in intra-plate basins. Occurrence of reserves, age and distribution of the sediments that generated or preserved them, and the formation of the mainly large structural (and other) traps are linked intimately to differing histories of plate margin evolution. The proper understanding of these histories could lead to additional reserves being established. The Arabian plate margins evolved at different times, the first being the northeast passive margin. This permitted the almost uninterrupted accumulation of thick sediments over a vast area including areally extensive organic-rich source rock deposits as well as good reservoir and seal units. The north/northeast margin(s) became a collisional boundary and a new Levant margin became a transform boundary in the Neogene.

Consolidation of the Afro-Arabian craton in the latest Proterozoic and Early Cambrian created a prominent north–south basement "grain" and a northwest–southeast (Najd) shear fracture system. Rejuvenations (affecting structures/sediment patterns) occurred in later periods and have controlled major hydrocarbon occurrences.

From latest Proterozoic to late Paleozoic time, the present north/northeast Arabian plate margin region, Anatolia, central Iran and the Afghan and Indian plates formed part of the long and very wide northern passive margin of Gondwana. This region was intermittently covered by shallow epeiric seas and bordering lowlands.

The present northeast margin formed as a result of Late Permian extension, rifting, and drifting away of former Gondwanan margin blocks and opening of the neo-Tethys. The oceanic domain was propagated further westward to form the northern passive margin by the Early Cretaceous. The southeast margin (Oman) had both a rift and a transform history from the (?Late) Jurassic. Ocean plate formation in the eastern Mediterranean (Early to Middle or ? Late Jurassic) and the Afro-Indian rift (Late Jurassic/Early Cretaceous) formed the northwest transform and the southeast passive margins of the Levant and Dhofar/Somalia. The Red Sea (western) and Gulf of Aden (southwest) passive margins were created as a result of Neogene extension, rifting, and propagation of the Indian Ocean spreading ridge westwards, separating Arabia from Africa and rotating and propelling the Arabian plate northeast to underthrust and suture with Eurasia. These motions created the new and current Levant ("Dead Sea") transform on the northwest plate margin of Arabia.

The northeast Arabian Mesozoic passive margin is characterized by the extraordinary width of its pre-erosional, carbonate and evaporite dominated, depositional shelf; reaching one and perhaps even two thousand kilometers. Across this broad shelf were interspersed intra-platform basins and swells; the latter either shed detrital material that intertongued as reservoir sediments with "basinal" source sediments or was deposited as neritic carbonate environments resulting in reservoir-quality strata. This paleogeography ensured an areally extensive distribution of multiple source/reservoir/seal units in the proper sequential order and optimum juxtapositional relationships for very efficient drainage of wide hydrocarbon kitchens into large, gentle, mainly structural closures. Leakage from these structures was minimized by the presence of effective regional seals. The Mesozoic depositional shelf along the northwestern and northern margins was relatively narrower with poorer areal distribution of source/reservoir deposits, further complicated by considerable deep removal of sedimentary section during Late

Jurassic/Early Cretaceous block movements. By contrast, the southeast margin remained relatively positive throughout most of the Mesozoic with the thin sedimentary cover leaving its Infracambrian/early Paleozoic source sediments at relatively shallow depths and consequently well within the liquid oil window.

The hydrocarbon potential/plays for each margin and for intra-plate basins differ according to their individual evolution history. The southeast margin is principally an Infracambrian-source/Paleozoic reservoir province with possibilities of subordinate Mesozoic or Tertiary plays in offshore basins that to date have yielded no discoveries. The northeastern margin is proven to be a large Mesozoic-sourced Mesozoic and Tertiary reservoir province. However, it has further potential for substantial new reserves generated from Silurian source rocks and trapped in Permian-Carboniferous sand (e.g. Central Saudi Arabia) and Late Permian carbonate reservoirs (Gulf region, Oman). In addition, the possibility of extensive Infracambrian source rocks charging Mesozoic (or younger) reservoirs though multistage migration via one or several paleotraps should not be ruled out, with the likelihood of older but less deeply buried Paleozoic reservoirs still containing oil.

The potential and play concepts of the northern/northwestern margins are less straightforward to categorize. In within-plate basins (aulacogens, grabens, intracratonic sags) Triassic and Late Cretaceous hydrocarbon sources are in the oil window and are increasingly found to have generated light oil/gas condensate. Toward the margin edge, and as a consequence of complexities in distribution due to Late Jurassic/Early Cretaceous deep erosion and Late Cretaceous and Miocene-Pliocene collisions, late Mesozoic source rocks preserved in troughs have either generated heavy sulphurous oil because of marginal maturity, or meteoric water flushing has been responsible for biodegradation. Silurian shales, however, both in margin-edge areas and within-plate basins of the northwest/northern regions where preserved, will prove increasingly to have generated hydrocarbons. However, future discoveries (e.g. as in southeastern Turkey, Syria, and Jordan) will require a better understanding of the Silurian subcrop distribution.

For the passive margins formed during the Neogene in the west (Red Sea) and south (Gulf of Aden), only minor discoveries have been made to date. Play targets for the former are the Neogene with good source/reservoirs/seals, and possible Jurassic objectives in the southern end of the margin. In the south the Paleogene with subordinate Neogene and Jurassic objectives appear to be viable plays. Organically rich, but immature Eocene source rocks are widely exposed on the Gulf of Aden rift shoulders, but can reach maturity offshore as a consequence of rapid subsidence and rift- related high heat flow. Within-plate grabens adjacent to the southwestern margin already have important Jurassic reserves and production (Ma'rib-Shabwa), and other grabens currently are being explored.

Arabian within-plate structural basins include failed rifts, half grabens, intracratonic sags, and strike-slip pull-apart/compressional belts. Arabian margin basin structuring includes epicratonic basins and overthrust belts. The Palmyride belt (Syria), the Ma'rib-Shabwa grabens (Yemen), the Risha basin (northeast Jordan/Iraq), the Rub al Khali basin (Arabia) and the thrust belts of the Taurus, Zagros and Oman mountain fronts show varied exploration promise and are proving to have additional hydrocarbon reserves. The Dead Sea-Levant Fracture basin(s) potential remains unassessed but locally appears promising.

Chapter 1

Middle East Hydrocarbons in 1990

The published figures of recoverable petroleum reserves for the Arabian plate region of the Middle East (Figure 1 and inset) at the start of 1991 were reported to be 663.226 billion (B) bbl of oil and 1,325.40 trillion cubic feet (tcf) of gas (88.20 B tonnes and 37.53 T cu.m respectively) (Oil and Gas Journal, 31 Dec 1990). This is equivalent to 66.4% of the world's proven recoverable oil reserves and 31.5% of the proven recoverable gas reserves. These figures are significantly higher than those published for the start of 1987, when the recoverable reserve figures given then were 402.28 B bbl of oil and 925.76 tcf of gas, or 57.6% and 25.5% respectively of the world's recoverable oil and gas reserves (Beydoun, 1988). This substantial increase is due mainly to the virtual doubling of the quoted Iraq and Iran oil reserves, the tripling of the Abu Dhabi oil reserves, and the increase of Saudi Arabia's published oil reserves by about 100 B bbl, together with more modest increases in other states in the area and to some new discoveries in previously non-producing states (Table 1). A major part of the verified increases is due to new field discoveries or to new deeper pool discoveries in established fields (particularly with regard to increases in gas reserves). Part of the increase may also be attributed to the reassessment of existing field parameters, including reservoir areal extents and improved recovery factors. However, some of the increases claimed by countries that have not carried out any noticeable new exploration may be due mainly to international political competition that causes them to feel compelled to raise their "proved" reserves. Nevertheless, genuine large increases have occurred that include substantial new discoveries. The enormous reserve total is attributable to 275 producing fields and at least 285 new and/or relatively new but as yet nonproducing or unassessed discoveries (including gas), and to deeper pools in established fields (Enclosure 1). Another 55 unidentified but reported discoveries may be added to the total. The figures show an increase over those quoted for end 1986 (Beydoun, 1988) by 20% and 12%, respectively, and indicate that in most of the countries active exploration and development drilling has been going on steadily and quietly.

Indeed, if reports from Iraq are typical, most of these increases result from new or deeper pool discoveries in the majority of the countries of the area. In Iraq, exploration in the period 1970–1989 involved extensive seismic surveys with supplementary magnetic and gravity surveys. Fifty-two structures were drilled, which led to the discovery of 38 new oil fields, and new productive reservoirs were established in 30 other known fields. The current Iraqi 100 B bbl recoverable oil reserves figure is distributed among 73 oil fields of which six are in the super-giant (>5 B bbl) category and 23 are in the giant (>1 B bbl) category; 76% of the reserves are in Cretaceous reservoirs, 23.9% in Tertiary reservoirs and only 0.1% (so far) are in Jurassic/Triassic reservoirs (Al-Khirsan and Al-Siddiki, 1989). These authors also stated that the surface and seismic surveys indicated the presence of 400 closed structures of which only 114 have been drilled, while in two-thirds of the known fields, exploration drilling had only been done as deep as to Cretaceous reservoirs.

In Saudi Arabia, very important if not major new reserves have been established over the last year in Paleozoic reservoirs that previously had never been seriously explored, and in an area previously undrilled outside the old Aramco concession. It should be added that brief reports of new discoveries have been released within the old Aramco concession. In the U.A.E., particularly Abu Dhabi, many new (smaller) fields have been discovered on small closures sandwiched between the giant and super giant fields discovered earlier.

The increase in the number of oil fields reported as being actually in production at end of 1990 (Oil and Gas Journal, 31 Dec 1990) is due mainly to new smaller-yield fields in southeastern Turkey, southern Oman and, to a lesser extent, in Syria; and also to the addition of Yemen as a multi-field producer. However, the producing-fields figure reported in that issue of the *Oil and Gas Journal* for Saudi Arabia seems far above the actual; it includes many long-discovered fields that are not yet connected to the gathering systems. Total Arabian plate field production averaged 16.635 MMBO/day in 1989, which was over 25% greater than the 1986 figure (12.365 MMBO/day) and amounted to 27.5% of total world oil production, compared with 22.1% in 1986 (Oil and Gas Journal, 31 Dec 1990; Beydoun, 1988).

The Arabian plate today is defined by plate boundaries formed in the early Mesozoic (the northeast, southeast, north and older northwest margins) or in the middle to late Tertiary (the south, southwest, modified northeast and north, and new northwest margins). The plate history that has controlled the hydrocarbon geology falls into four stages:

1. The continental basement was formed as an eastern part of the Nubo-Arabian craton during the late Proterozoic.

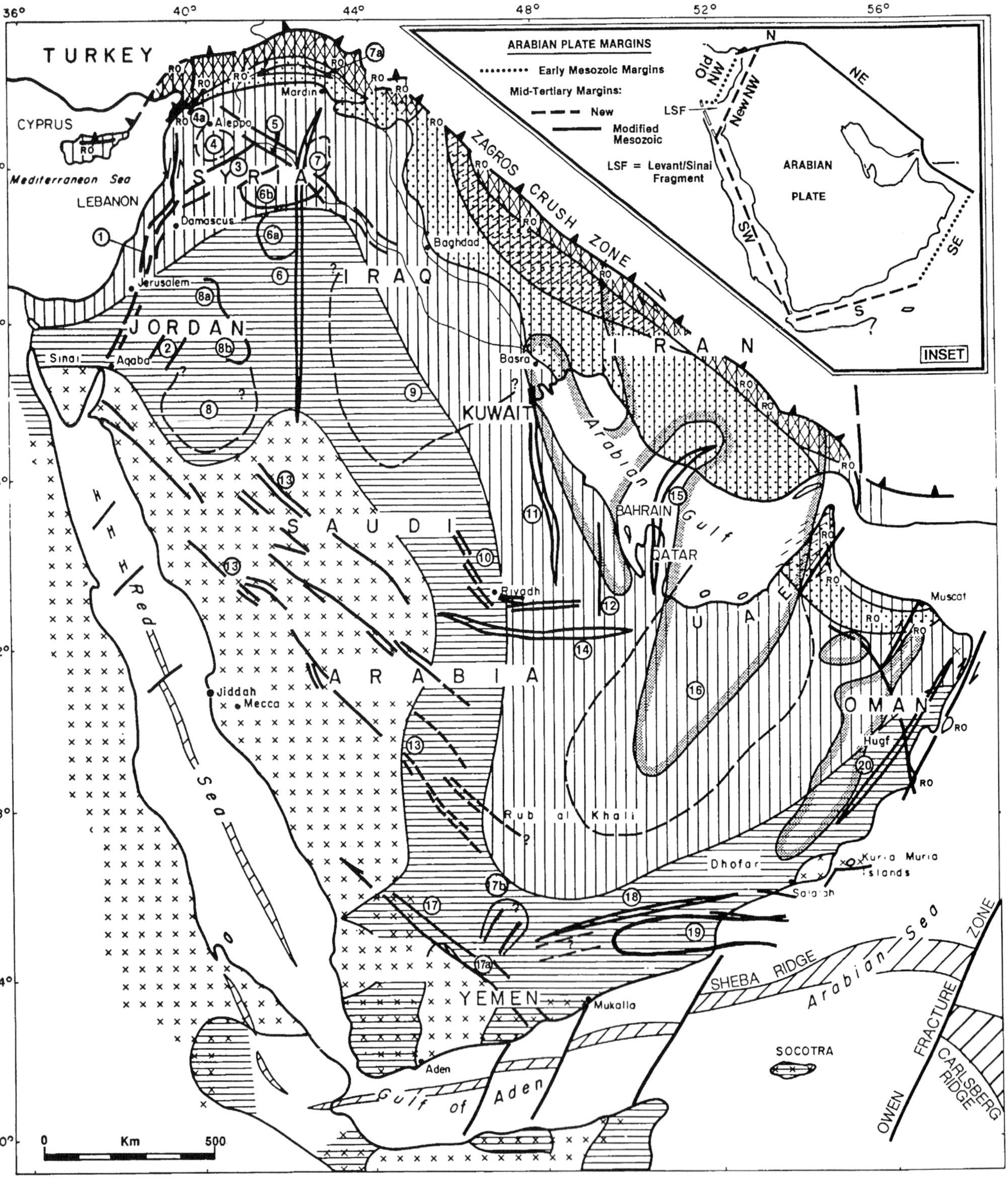

Figure 1—Tectonic framework sketch map of the Arabian plate. Adapted and expanded from Beydoun (1989) after Henson (1951). Reproduced with permission of OAPEC.

2. The region was part of the marginal shelf of the Gondwana supercontinent through the entire Paleozoic until the Permian.

3. The end of the Paleozoic and the early Mesozoic plate divergence left a continental "Arabian Promontory" (e.g., see Figure 25) as a spur from the African plate.

4. A middle to late Tertiary collision with "Eurasia" compressed and modified the north and northeast

LEGEND

- Arabo-Nubian Shield, and with some Mesozoic – Cenozioc Cover of Stable Shelf
- Stable Shelf
- Unstable Shelf
- Zone of Upper Cretaceous Marginal Flysch Troughs
- Zone of Neogene Marginal Troughs and Autochthonous Folding
- Zone of Thrusting and/or Gravity Nappes
- Magnetic Zones (New Oceanic Crust)
- RO Radiolarite/Ophiolite Complexes (U. Cretaceous mainly)
- Principal Fault Zones
- Infra-Cambian to Early Cambrian Salt Basins

PRINCIPAL STRUCTURAL FEATURES

1. Levant Fracture System
2. Jafr 'Basin'
3. Palmyride 'Basin' (Aulacogen)
4. Aleppo High (4a) Aafrine Basin
5. Euphrates Graben
6. Hail-Jauf-Ga'ara-Mardin Arch (Swell)
(6a) Risha "Basin" (6b) Rutbah High
7. Khleisia High (7a) Mardin-Buzova Swell
8. Tabuk Basin (8a) Azraq Basin (8b) Sirhan Basin
9. Widyan Basin
10. Central Arabian Graben and Trough System
11. An-N'ala Axis
12. Ghawar Axis
13. Najd Fault System
14. Central Arabian Arch
15. Qatar-Fars Arch
16. Rub-al-Khali Basin
17. Jawf-Marib Graben (17a) Shabwa-Balhaf Graben (17b) Sir Basin
18. Hadhramaut Arch
19. Jeza'a Trough
20. Huqf-Ja'alan Axis

margins, whilst new rifting divided the Arabian plate from Africa to form the south, southwest and new northwest margins. (Figure 1, inset).

However, within the Arabian plate the overwhelming proportion of the proved recoverable hydrocarbon reserves is located along the northeastern Arabian plate margin within the Iraq-Iran-Arabia basin, which stretches from northwest Iraq through southwest Iran/eastern Arabia/the Gulf, to north-central Oman. This huge, elongate basin contains 98.7% of the Arabian plate's recoverable oil reserves and 97.2% of its recoverable gas reserves.

The occurrence of these vast, established hydrocarbon reserves in the northeast plate margin area, along with the relatively minor amounts along the other margins and in within-plate basins, is intimately linked to the individual plate margin and intra-plate basin evolutionary history. Indeed, the potential for further new discoveries both from currently known Mesozoic/Cenozoic reservoirs and from older, stratigraphically and physically deeper reservoirs, should also be considered in the same plate tectonic context. This is especially true for understanding the factors that controlled the distribution of the sediments that sourced and preserved these reserves and gave rise to the trap-forming mechanisms. The proper analysis and comparison of the differing geological histories of each of these distinct plate areas will accelerate the finding of new reserves in the main northeast margin basin, and will also enhance chances of discoveries in the less-productive plate margins or in such relatively sparsely explored intra-plate basins as the Palmyrides basin and Euphrates graben of Syria/Iraq, the Marib-Jauf-Shabwa-Balhaf basin of Yemen, and other comparable basins within the Arabian plate.

In the following sections, an up-to-date outline review of each Arabian plate margin (and principal within-plate basins) evolution based on recent published results is given.

TABLE 1: Arabian plate (Middle East) reported proved published oil and gas reserves per country at end 1986 and end 1990.

COUNTRY	OIL RESERVES (Billion bbl) (1) 1986	(2) 1990	GAS RESERVES (tcf) (1) 1986	(2) 1990
Abu Dhabi (UAE)	31.000	92.200	90.000	182.800
Bahrain	0.140	0.097	7.000	6.250
Dubai (UAE)	1.350	4.000	4.400	4.800
Iran	48.800	92.850	450.000	600.350
Iraq	47.100	100.000	28.000	95.000
Jordan	Unassessed	0.020	Unassessed	.400
Kuwait	91.916	94.525	35.000	48.600
Neutral Zone	5.212	5.000	12.400	10.000
Oman	4.032	4.300	7.500	7.200
Qatar	3.154	4.500	152.000	163.200
Ras al Khaima (UAE)	0.200	0.400	1.200	2.000
Saudi Arabia	166.573	257.504	123.990	180.355
Sharjah (UAE)	0.500	1.500	9.000	10.800
Syria	1.400	1.700	4.200	5.500
Turkey	0.400	0.650	.460	1.150
Yemen	0.500	4.000	.600	7.000
TOTAL:	402.277	663.226	925.760	1,325.405

Sources: (1) Beydoun, 1988 (based on Oil and Gas Journal)
(2) Oil and Gas Journal, 31 Dec 1990.

Chapter 2

Cratonization of the Arabian Shield: The Basement Grain is Imprinted

The evolution and consolidation of the Arabian craton has been the subject of considerable investigation and research over the past two decades most of which is outside the scope of this book. However, the early evolution created prominent basement grain directions and shear fracture systems (most notably on north–south and northwest–southeast trends) that later had varied effects on younger structuring and sediment patterns. This in turn controlled major hydrocarbon occurrences. Therefore, a brief summary is essential to understanding hydrocarbon occurrence.

The evolution of the Arabian basement and consolidation of the Arabian shield have been discussed by many authors whose work was reviewed by Beydoun (1988, 1989b). Stoesser and Camp (1985) and Brown et al. (1989) provided excellent recent succinct summaries. Husseini (1988, 1989) also provided recent reviews that addressed some of the contradictions in different basin evolution models and attempted to reconcile them.

The primary crust of the western Arabian shield is composed of a combination of several constructional units, each of which is an intra-oceanic island arc terrane consisting of an andesitic assemblage of metavolcanic rocks and a dioritic suite of plutonic rocks. There are at least three such terranes. The ensimatic island arc development and progressive cratonization by coalescence of arc terranes took place over a long time span that began about 950 Ma with formation of the neocraton by accretion of arcs. Each closure and arc collision led to deformation and ophiolite obduction now preserved as cryptic sutures that segment the shield in approximately north–south units and imply a general east–west convergence. This process culminated with microplate and (?micro) continental collision (preserved as small continental affinity terranes in the eastern shield) at about 640 Ma, which Brown et al. (1989) have termed the culminant orogeny.

The subsequent crustal history is characterized by growth of elongated north-trending gneiss domes and other similar-trending structures. These and other intracratonic magmatic intrusions and tectonism are also collision related. They, as well as post-tectonic granite intrusions, followed a north–south orientation parallel to the trends of the old structural grain. Much of the north–south grain was subsequently offset by the northwest–southeast-trending, latest Proterozoic-earliest Paleozoic, strike-slip Najd fault system. The changing sense of motion on this system is related partly to the combined effects of east–west collisional event(s) (Stacey and Agar, 1985) and probably in part also to crustal extension in northeasternmost Afro-Arabia (Stern, 1985; Husseini, 1988). The rift-related and/or collision-related events, including the conversion of the Najd fault system from a dextral (Stacey and Agar, 1985) to a sinistral strike-slip fracture system, spanned the period 640–550 Ma. The Najd system is clearly defined on the surface of the Arabian shield over a width of 300 km and a strike length of more than 1,200 km (Figure 1). Brown et al. (1989) consider this faulting event as the final tectonism/cratonization event of the Arabian shield that was caused by east–west convergent forces fracturing the now thick shield along a number of great shears. The left-lateral northwest system (the Najd trend) prevailed over the conjugate dextral northeast-trending fractures (or over the extensional northeast-rifting phase faults of northernmost Afro-Arabia). However, both northwest and northeast trends can be seen in some parts of the Arabian plate outside the shield area, particularly to the north and southeast, where they have been subjected to later rejuvenation by new forces linked to plate margin evolution.

Whatever the earlier history of Najd fault motion, the final phase is left lateral and the system consists of braided en echelon series of fractures with sinuous and branching geometry and a slight convergence to the northwest. This convergence resulted in the development of a number of local extensional segments with grabens filled with taphrogeosynclinal sediments and volcanic rocks—the J'balah or (now) Jubaylah group of Delfour (1970). Additionally, there are a number of ultracompressional segments with elongated gneiss domes, or where J'balah sediments were folded into asymmetric synclines during final phases of movement as cratonization was being completed.

The effects of Najd stresses are found throughout the shield and possibly many of the northeast-trending structures in the western shield bordering the present Red Sea may be of Najd age (Brown et al., 1989). With completion of cratonization at about 550 Ma and an end of activity on the Najd sinistral system, epicontinental subsidence developed differentially and resulted in the formation of broad Paleozoic basins.

The Najd faulting event had far-reaching repercussions not only in the shield area but also in the newly stabilized craton bordering it. Spanning, as it did, terminal events of the Proterozoic and the dawn of the Phanerozoic, the faulting was responsible for the creation of a number of rift-related basins and crustal sags on conjugate and other trends elsewhere. Greenwood et al. (1980) suggested that the Infracambrian/Early Cambrian salt basins ("rhombochasms") of southeast Oman are possible examples of such structures, spanning with their sediments the Precambrian-Phanerozoic boundary and including amongst these, widespread deposits of organic-rich source sediments and salt. The later movement and partial dissolution of this salt controlled much subsequent structuring in this region, which had a considerable effect on hydrocarbon generation, migration, and preservation. These rift and other basins collectively represent the first intraplate structural phase within the newly consolidated Arabian portion of Gondwana.

Chapter 3

Paleozoic Evolution of the Arabian Margin of Gondwana

From the latest Proterozoic to the Late Paleozoic, Arabia with Turkey (in several continental fragments?), central and northwest Iran, and also with the later Afghan, Indian and other plates, collectively formed part of the long and wide "northern" passive margin of Gondwana bordering the paleo-Tethys ocean (Figures 2, 3). Much of this region was covered intermittently by shallow epeiric seas that bordered lowlands made up of portions of the shields (Nubo-Arabian, Indian, etc.). The areal extent of the shelf seas changed in response to succeeding transgressions and regressions as the Paleozoic era advanced, and their setting varied between tropical and temperate latitudes of the southern hemisphere. Several migrations into high latitudes and polar proximity resulted in glaciations that affected parts of the region in the late Proterozoic, the Late Ordovician and the Late Carboniferous-Early Permian. The orientation of this passive margin rotated between northwesterly facing in the latest Precambrian with oscillations to a westerly setting in the Cambrian and back to northwesterly, then to northerly facing in the Devonian. The passive margin orientation swung to the northeast in the Carboniferous and was almost easterly facing in the Permian (Scotese et al, 1979; Scotese and McKerrow, 1990) (Figures 2, 3). However, there still is uncertainty and question regarding the position and numerical composition of the central Iran and Afghanistan blocks relevant to one another and to their position on the plate margin. Similarly, there is uncertainty about whether Turkey was adjacent to the Levant (as is already established by other workers for the early Mesozoic–see Figure 11A), or next to central Iran as shown in Figures 2 and 3. If Turkey was not next to central Iran, then the Iran fragment (Figure 2) could represent the Sanandaj/Sirjan block with northwest Iran and Central Iran aligned with it and located further out adjacent to Oman (Sengor et al 1988) (Figure 11A), but these data could still allow other interpretations (W.S. McKerrow, personal communication, 1990)[1]. Whatever the jigsaw fit selected, the width of the epeiric shelf along this margin reached at least 2,000 km and may have stretched to 3,000 km or more if the position of central Iran adjacent to southeast Arabia (the Gulf and Oman) is accepted, as stratigraphic correlation and facies patterns strongly indicate (eg., Stocklin, 1968, 1986; Beydoun 1988, 1989b) (Figure 5).

Climatically, based on the consensus of plate tectonic data and the predominance of clastic deposition in the Paleozoic era, the region was mainly temperate with episodes of warm water carbonate deposition in the Middle Cambrian, the Devonian and the Late Permian, and two proven glaciations in the Late Ordovician and Carboniferous-Permian, together with sedimentological evidence from Oman of a late Proterozoic glaciation. The generalized paleolatitude curve for Oman (Figure 4) reflects this quite well and at the same time also indicates some of the uncertainties in the data relating to the Paleozoic.[2]

With completion of cratonization and consolidation of the region by the end of the Najd faulting event, this vast area became largely stable, and throughout most of the Paleozoic, it underwent only periodic broad epeirogenesis with some more localized taphrogenesis. An extensive supply of mainly clastic sediments was deposited in a number of large basins, for which erosion of the bordering uplifts or intervening swells provided the dominant sediment.

The earliest sedimentary succession spans the Proterozoic–Phanerozoic boundary and was laid down in rift-related basins and grabens or in broad intra-shelf and epicratonic sags. This occurred in many parts of the region, but because of subsequent thick, younger sedimentation and deep burial, most of these deposits have seldom or ever been penetrated by the drill. They are most completely known from areas that escaped deep burial and where subsequent tectonism has brought them closer or to the surface, especially in the

[1] Interestingly enough, in as yet unpublished plate reconstructions for the Ordovician, Turkey is placed next to the Levant and what are marked as Turkey and Iran on Figs. 2 and 3 here, are substituted by a small Hainan block and a much bigger South China block (Scotese and McKerrow, in press). This only highlights the still inherent uncertainties in the earlier plate reconstructions, which have variably reliable control data points, which can exhibit important differences depending on the data parameters and data sources used.

[2] Revision of the Silurian to mid-Carboniferous data in 1989 brought out an unexpected drift to tropical/equatorial latitudes. Using the Scotese and McKerrow (1990) position of Muscat in Fig. 3 over the same time interval, a much more subdued but smooth curve about halfway between the 1982 and 1989 curves of Fig. 4 emerges. Clearly, the various discrepancies indicated are artifacts of the data available and of the parameter selections, and suggest caution in overly detailed conclusions when using the various plate reconstructions at hand.

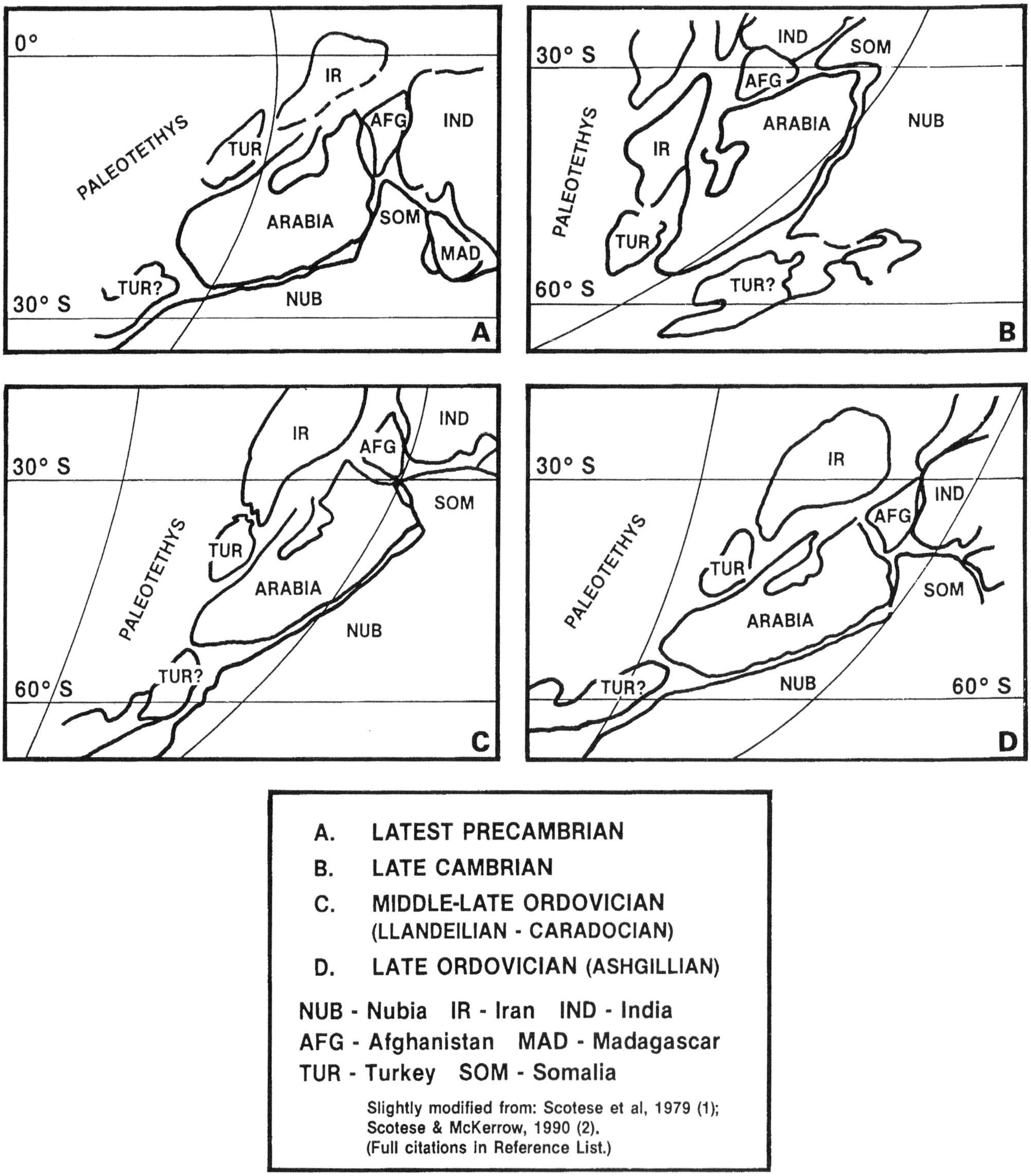

Figure 2—Early Paleozoic plate tectonic reconstructions: Arabia and surroundings.

extensive outcrops of the Huqf-Haushi swell of Oman and in the subsurface of southeast Oman. Deposits of similar age (and often similar facies) are present in patchy outcrops in southwest Iran and southeast Turkey as a result of vertical uplift during later continental collision and thrusting. Similar deposits are known from other parts of the Arabian plate including Yemen, the Arabian shield, southern Jordan, northern Somalia, and Egypt. Correlation of the main sequences found along the plate margins and in intra-plate basins are given in Figures 5 and 6.

The Oman deposits, collectively known as the Huqf

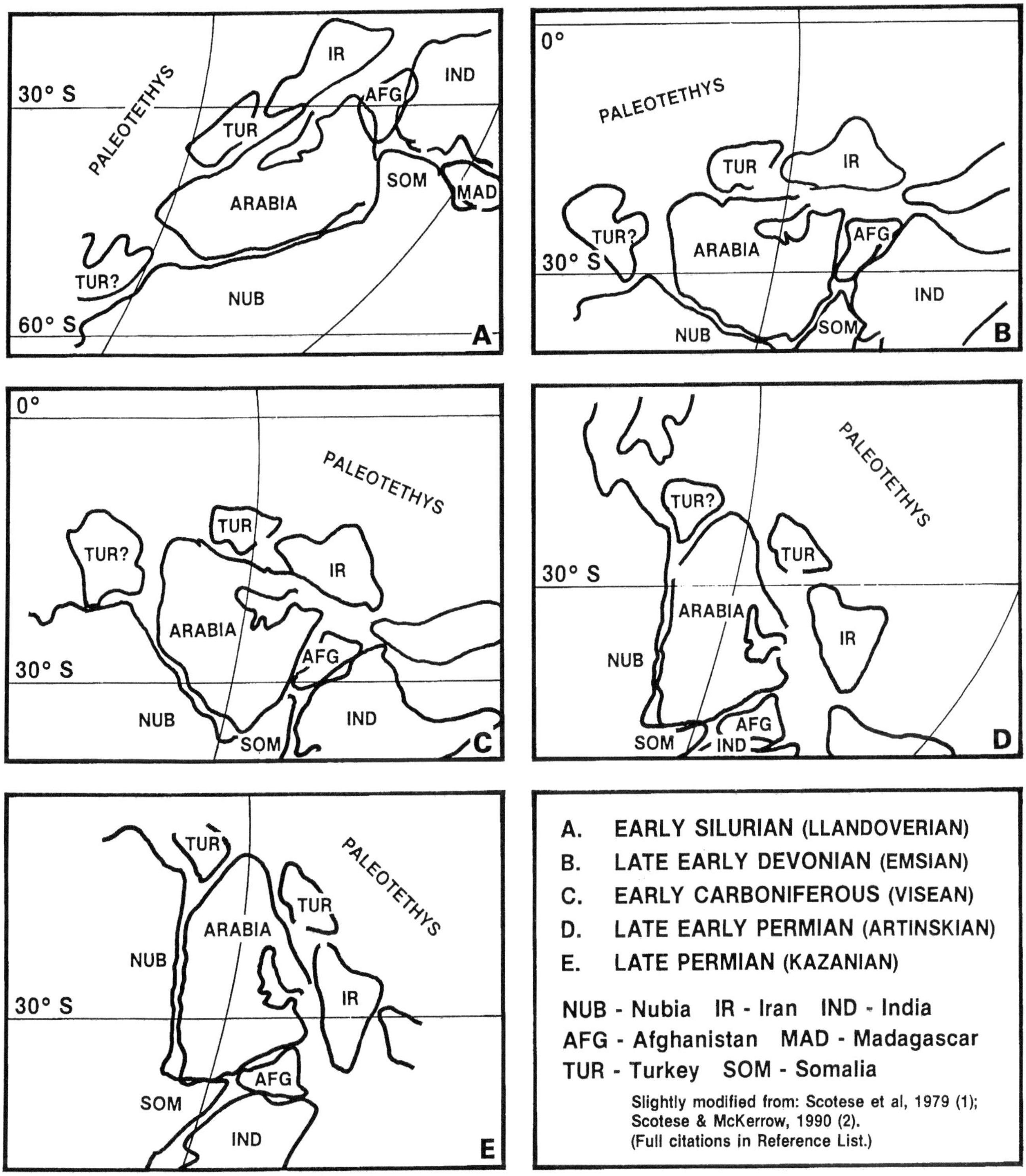

Figure 3—Later Paleozoic plate tectonic reconstructions: Arabia and surroundings.

group, consist of a thick succession of sediments that form two clastic formations alternating with two carbonate formations and capped by a terminal evaporitic formation with carbonate intervals. The evaporitic Ara formation may be in part laterally equivalent with the underlying uppermost carbonate Buah formation. Gorin et al. (1982) summarized the main characteristics of the individual formations and correlated each with an equivalent formation exposed in the core of the Oman Mountains. The depositional environments represented by the various formations are shallow marine (subtidal) to tidal flat and bordering lowland regimes.

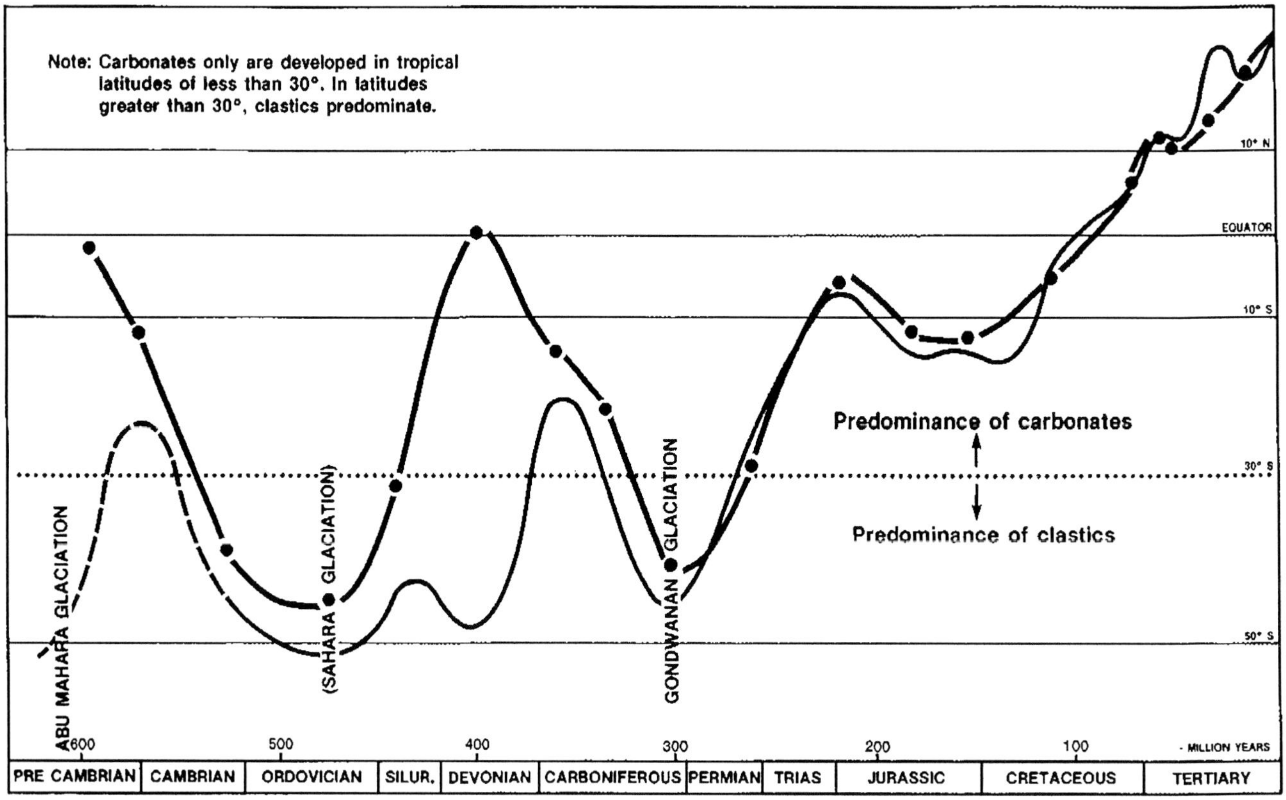

Figure 4—Generalized palaeolatitude curve for Oman.

There are also occasional lagoonal conditions that extend into low-energy, quiet marine, with the terminal mainly evaporitic sequence deposited under restricted super-saline conditions in basins bordered by a carbonate-anhydrite platform. Wright et al. (1990) concluded that the two carbonate formations were deposited as prograding fill sequences beginning below wave base in intraplatformal basins.

The age of the Huqf group is taken as late Proterozoic Vendian to Early Cambrian (Gorin et al., 1982; Beydoun 1988, 1989b; Mattes and Conway Morris, 1990), and the evaporitic Ara formation is considered correlative with the Hormuz salt of Iran. In Iran, there is more than one evaporitic cycle in the Hormuz complex and its age is late Proterozoic to Early Cambrian (Stocklin, 1986). For a number of reasons discussed elsewhere (Beydoun, 1989b), the Ara salt of Oman is here inferred to correlate with the younger Hormuz evaporitic cycle and is considered to be of Early Cambrian age (see also Husseini, 1988, 1989) (Figure 5). This view differs from that of Hughes Clarke (1988), who equates the Amin formation (basal Upper Haima group) of Oman with the Lalun Lower Cambrian formation of Iran (Stocklin, 1986), which implies that the underlying Ara salt equates with the older Hormuz evaporites.

The Buah-Ara cycle in Oman is unconformably overlain by clastics that are assigned Early Cambrian to Early Silurian ages (Lower and Upper Haima groups: Hughes Clarke, 1988). Good oil source rocks occur in several formations of the Huqf group, particularly in parts of the upper pre-salt units (Shuram and Buah formations) as well as within the Ara evaporites (Gorin et al., 1982; Al-Marjeby and Nash, 1986; Grantham et al., 1988). Rich source rocks of Huqf group equivalents apparently were deposited over extensive areas of this wide plate margin, particularly in those extensional areas with substantial evaporitic development in the Persian Gulf region and borders (Figure 7).

The equivalent J'balah (Jubaylah) and Fatimah group deposits of the shield area include algal beds and shales but no evaporites. They have been identified from scattered boreholes and seismic data in the eastern Arabian subsurface east of the shield, but their distribution and facies are very poorly understood there. The equivalent deposits of southeast Turkey and Jordan are largely made up of sandstones or coarse clastics and organic-rich source rock beds appear to be largely absent. The clastic-carbonate group similar to the Huqf exposed in Yemen (Ghabar group) exhibits similarities to the J'balah group both in the presence of algal beds and also in the presence of volcanogenic clastics, as well as in having undergone incipient metamorphism (Figures 5 and 6).

Throughout the Arabian plate region, the unconformably overlying Paleozoic deposits are predominantly clastic. They are known from outcrops around

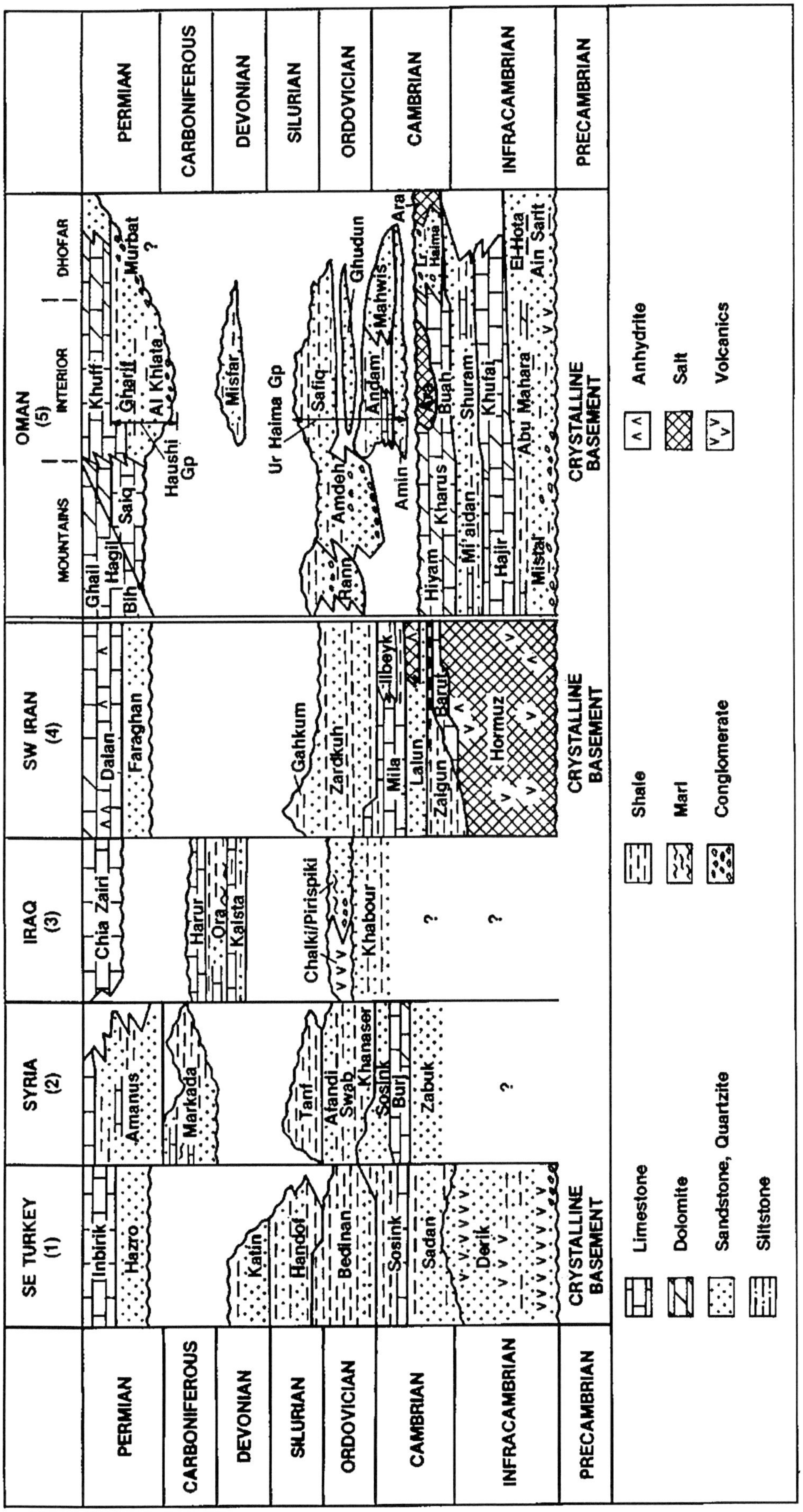

Figure 5—Infracambrian/Paleozoic formations correlation: southeast Turkey/Oman.

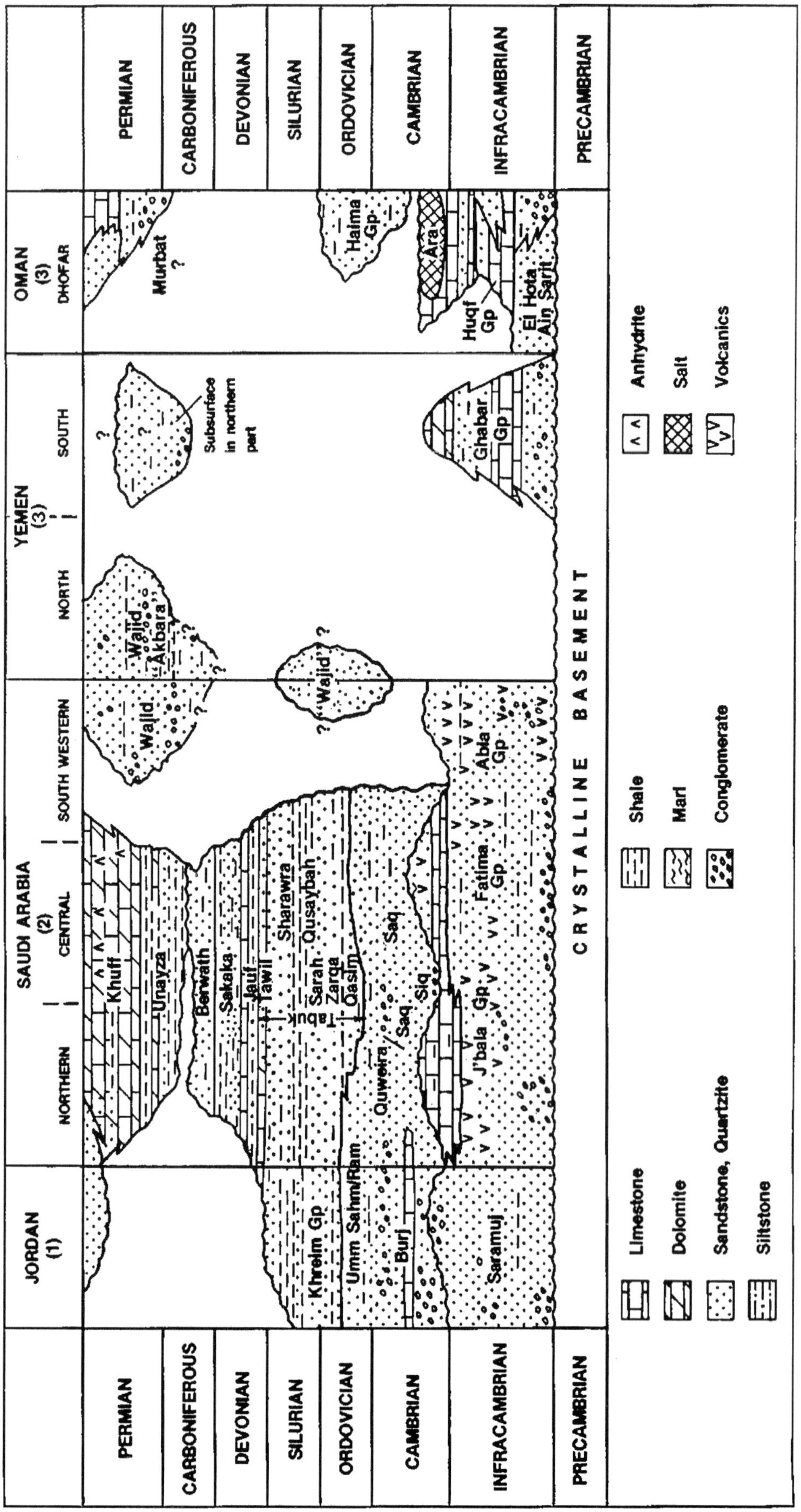

Figure 6—Infracambrian/Paleozoic formations correlation: Jordan-Yemen and Dhofar.

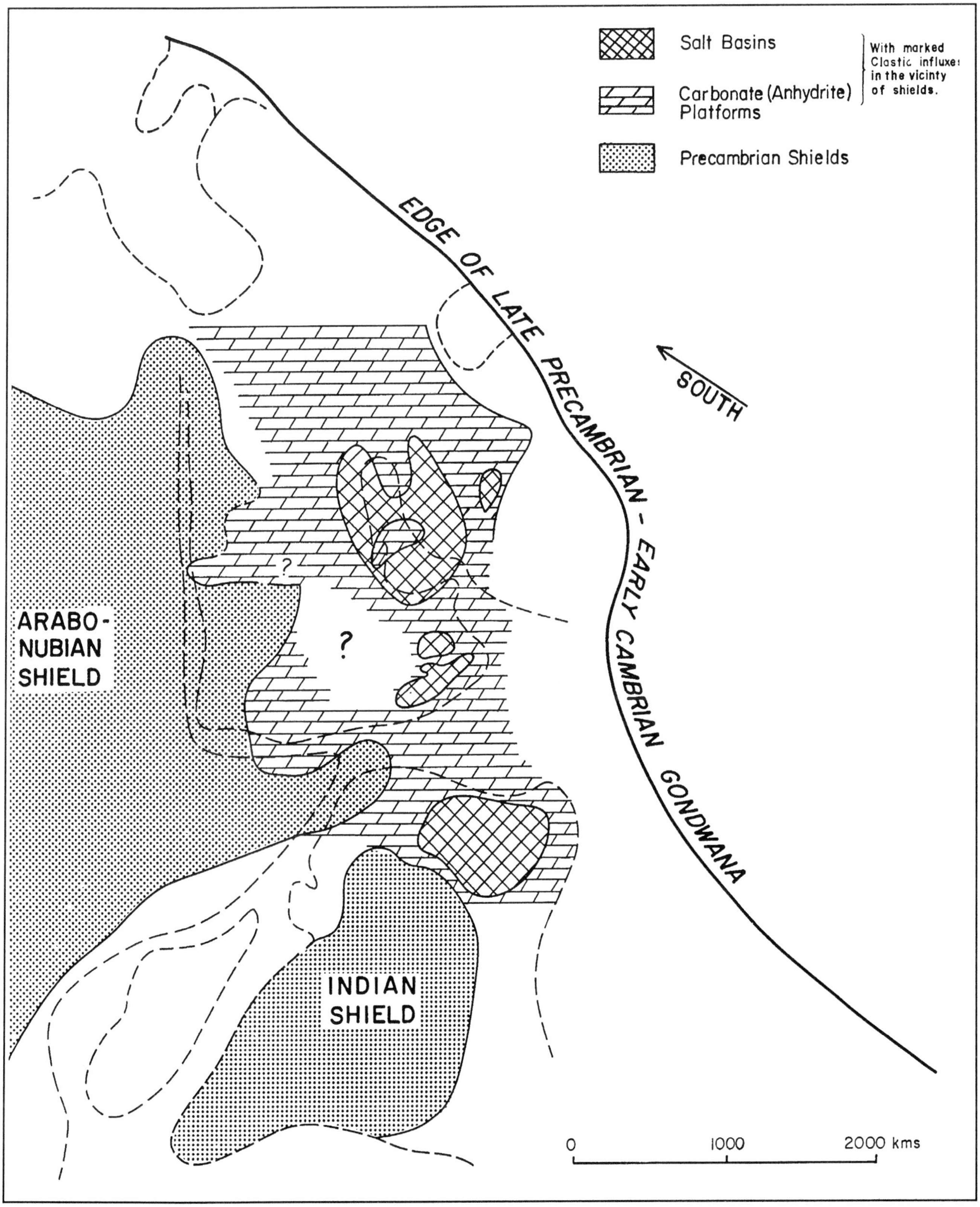

Figure 7—Paleogeography of the Late Precambrian/Early Cambrian sediments of Gondwana.

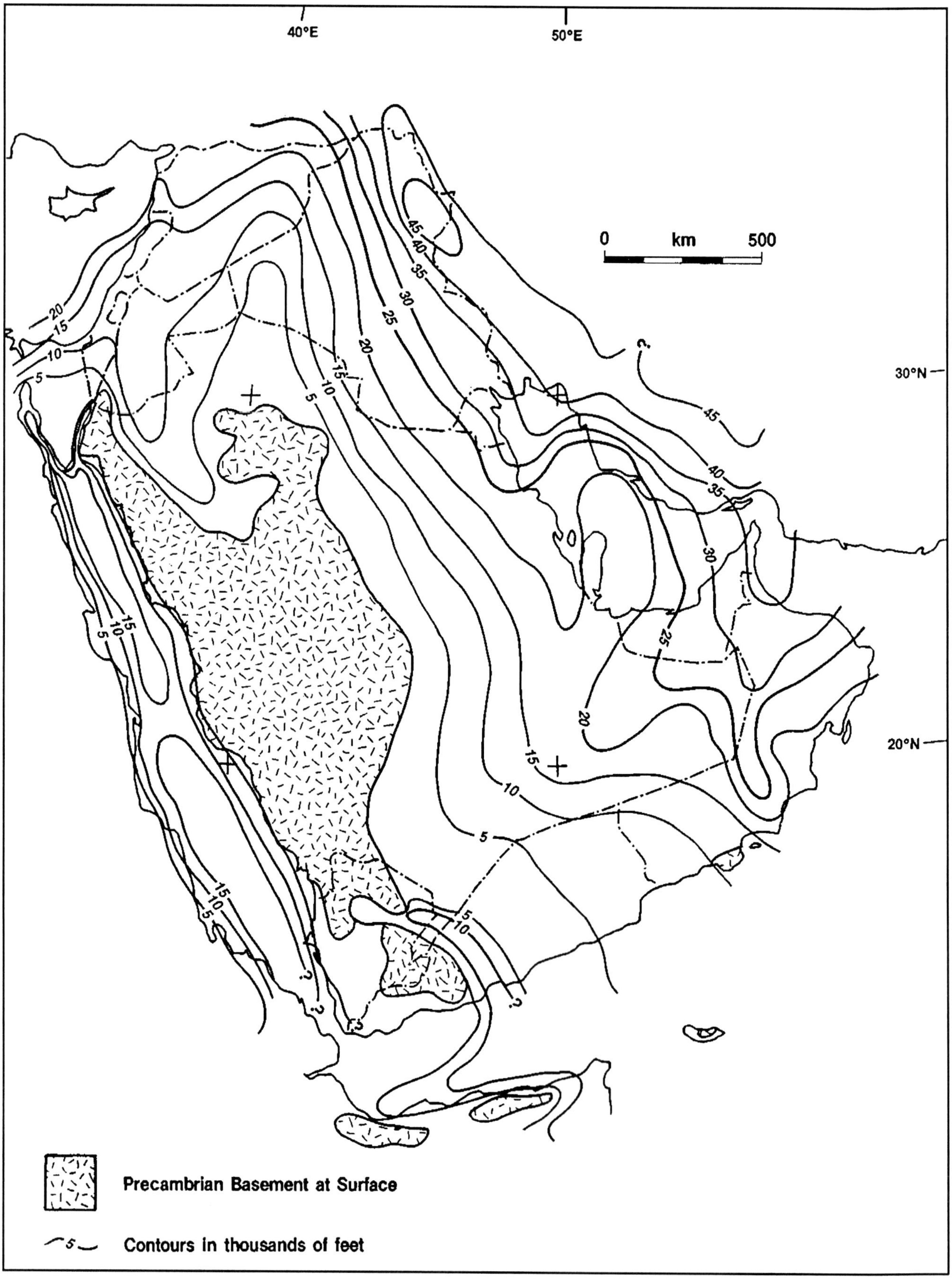

Figure 8—Arabian plate; approximate thickness of sedimentary cover.

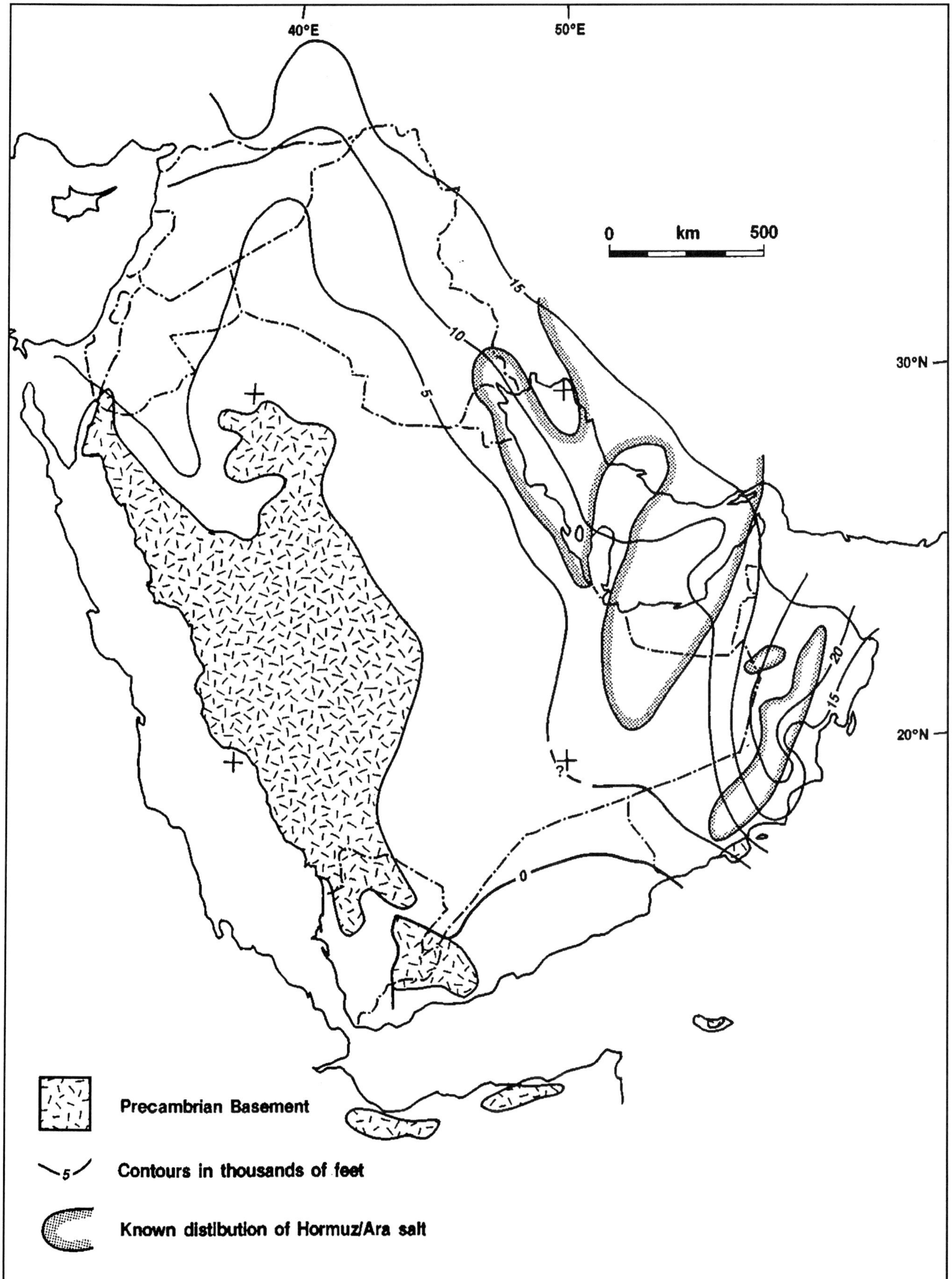

Figure 9—Arabian plate; approximate thickness of Paleozoic sediments.

the Arabian shield from southern Jordan through central and southwest Arabia (Yemen). They are also known from scattered outcrops along the northern and northwestern and northeastern Arabian plate margins in the Amanus-Taurus-Zagros and Oman fold belts of southeast Turkey, northwest Iraq–southwest Iran, and the Oman Mountains as well as the Huqf–Haushi axis, where younger tectonism has partly brought them to the surface and deformed and/or partly eroded them (Figures 5 and 6). Paleozoic strata are also exposed in scattered outcrops in the blocks constituting Central Iran, Turkey (Anatolia) and Afghanistan that had separated from the Gondwanan margin at the end of the Paleozoic with the opening of the neo-Tethys and resutured with the Arabian plate on closure of that ocean in the Neogene. However, the Paleozoic successions of the region between the shield-bordering outcrops and the Taurus–Zagros–Oman fold belts along the Arabian plate margin lie deeply hidden under thick, younger cover and have seldom been fully penetrated by the drill during petroleum exploration (Figure 8). Thus, details of the thickness and facies variations for the entire succession are poorly understood. Nevertheless, enough is known to confirm that the predominant Paleozoic facies are siliciclastic, with more temporarily limited episodes of carbonate and shale deposition, some of which, however, were also areally extensive. These various lithofacies for the Arabian plate margin region and elsewhere within the Arabian plate are generalized in the correlation diagrams shown in Figures 5 and 6. Along the north/northeast/southeast Arabian plate margins, total Paleozoic sediment thickness may reach several thousand meters (up to 15,000 ft) (Figure 9).

Carbonate depositional episodes occurred in Early Cambrian and/or Middle Cambrian time and are represented by the terminal Huqf group Buah and Ara carbonates, the carbonates of the J'balah/ Fatimah groups of Saudi Arabia, the Barut and younger Mila formations carbonates of central and southwest Iran, the Sadan/Sosink of southeast Turkey (Early–Middle Cambrian) and the Burj formation of Jordan and parts of Syria and northernmost Saudi Arabia (Early–Middle Cambrian) (Figures 5 and 6). The return to areally extensive carbonate deposition did not occur until the Devonian when the region was again in tropical latitudes (Figure 3 and 4). These are represented by the cyclical carbonates of the Jauf formation of Saudi Arabia (Early Devonian) and the Dadas (Katin equivalent) of southeast Turkey and similar carbonate deposits of the Khaista formation of Iraq (Late Devonian) and into southwest Iran/central Iran (Sibzar formation). Elsewhere, clastics laterally replace the carbonates or the carbonates are eroded. A subsurface carbonate occurrence in parts of Syria (Markada formation) inferred to be of Carboniferous age (Lababidi and Hamdan, 1985) may also be of equivalent Devonian age. Conceptually, however, it would be in keeping with paleolatitudinal considerations that carbonate deposition occurred in the Early Carboniferous and was subsequently mostly removed by uplift and erosion during Late Carboniferous–Early Permian episodes linked to glaciation elsewhere in the region, and/or by subsequent (Mesozoic) deep erosion phases.

The most areally extensive (preserved) carbonate episode occurred in the Late Permian closely following Gondwanan deglaciation that had affected the southern part of the Arabian plate and adjacent plates. This carbonate deposition is related to the rifting event that eventually separated off central Iran and other proximal fragments and opened the neo-Tethys ocean. The return to more tropical latitudes accompanying the deglaciation and rifting was responsible for the extensive warm-water Late Permian (to Early Triassic) transgression along the entire newly formed northeast passive margin and for deposition of extensive shelf carbonates constituting the Khuff formation and equivalents (Dalan, Chia Zairi, Inbirik, of southwest Iran–Iraq–southeast Turkey) (Figures 5 and 6). The middle and upper part of the formation includes interlayered subordinate evaporites, especially in the Gulf region (Figures 5 and 16) where this facies extends into the Triassic. The Khuff and equivalent formations constitute a major hydrocarbon reservoir in the Gulf region.

A significant facies development of potentially major importance to the hydrocarbon prospects of Paleozoic reservoirs of the Tethyan Arabian-plate margin(s) and intra-plate basins is the widespread deposition of organic-rich shales during the latest Ordovician–Early Silurian. This deposition occurred over the greater part of the wide Arabian epeiric shelf and is associated with several cycles of global sea level rise and transgression in a temperate climate. Representative sections of this shale deposition are to be found in the Safiq formation of Oman, the Gahkum formation of southwest Iran, the Tanf formation of Syria and the Bedinan formation of southeast Turkey along the plate margin; also in the outcrop sections around the Arabian shield (Qusaiba formation) as well as in the surface outcrops of southern Jordan and the subsurface of the northern and eastern part of that country (Khraim group). However, removal of this once widespread shale, either as a result of late Paleozoic epeirogenesis or even of later block movements associated with Mesozoic northern plate margin rifting and/or terminal Jurassic uplift (which led to very deep erosion of most of the uplifted blocks), has made the subsurface distribution of these potentially viable source rocks difficult to decipher. For example, along the well exposed Zagros belt they are totally absent in outcrop in Iraq and only patchily present in southwest Iran so that, while the overall possible occurrence can be outlined, the true subcrop presence can only be conjectural (Figure 10).

Evidence for the late Proterozoic glaciation (Abu Mahara/Mistal or basal Huqf group formations) is based only on the occurrence of glaciogenic diamictites in these formations both in outcrop in the Oman Mountains and in the subsurface of South Oman (Gorin et al., 1982) and a coincidence with well documented glaciations of this time from other parts of Gondwana. By the end of Huqf deposition, however, the region had drifted to lower latitudes with warm and arid conditions

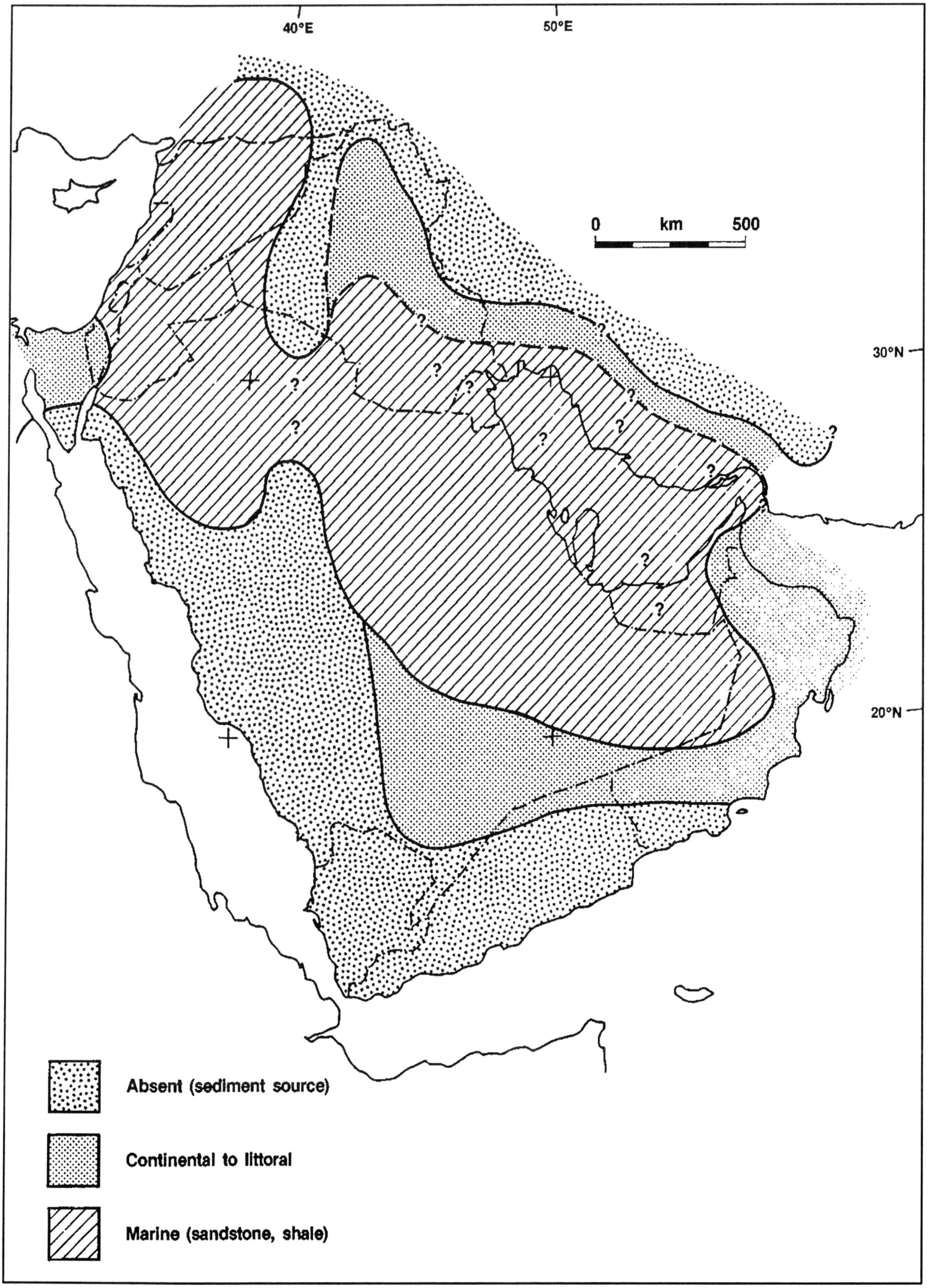

Figure 10—Arabian plate; Late Ordovician-Early Silurian probable distribution of facies.

prevailing by the advent of the Buah-Ara cycle (latest Proterozoic–Early Cambrian). These conditions tie in well with equivalent evaporites of the Indian plate, then in a proximal position (Figure 7).

Well-documented evidence for the Late Ordovician Saharan glaciation has been known for some time. Recent work in central Arabia (McClure, 1978) has revealed the presence of glaciated boulders and tillites, dated as Late Ordovician, with which proglacial sandstones interfinger laterally. If the paleolatidunal position of Arabia as given (Figure 2D) is valid, then Arabia would have been just within the margin of the ice-cap or perhaps the glaciation here, as possibly also in Turkey, may have been montane.

The Late Carboniferous–Early Permian glaciation covered only the present southern margins of the Arabian plate and there is well-documented evidence for this. Ice sheets extended from Ethiopia to southern Oman, and boulder beds of glacio-marine origin, as well as tillites, occur extensively and were first reported by Morton (1959), Helal (1964), and Dow et al. (1971). More recent field and subsurface work by de la Grandville (1982), Braakman et al. (1982), Qidwai et al. (1988), McClure, (1980), Kruck and Thiele (1983), El-Nakhal, (1984), and others, has confirmed the wide extent of these deposits and has linked them to similar deposits in the Salt Range of the Indian plate. However, in view of the paleolatitudinal position of the region (Figure 3D), the glaciation may have been partly montane or on the very edge of the Gondwanan ice sheet. In Oman, these glacial-related clastics can form excellent although laterally variable paired reservoir/seal beds that are ideal for hydrocarbon accumulations (Levell et al. 1988).

The sediment sources for the various Paleozoic clastics of the Arabian plate (and adjacent fragments/plates) that, in the main, were deposited in extensive, shallow, epicontinental seas, were the Arabian and Indian, etc. shields and other smaller positive features (Ha'il-Ga'ara Arch, the Hadhramaut Arch, the Huqf-Ja'alan axis, and others) (Figure 1). During subsequent emergence and erosion, cannibalistic reworking of clastics deposited earlier provided substantial sources for younger units on both a local and a more regional scale.

Evidence for any major Paleozoic tectonic movements in the region, other than for broad subsidence and uplift of an epeirogenic nature and of more localized taphrogenic block movements, is either sparse or lacking. The general absence of Devonian deposits from much of the region except for northwest Arabia, and the reduction of Silurian sequences or their removal from much of the Iranian portion of the Arabian plate and the Ha'il-Ga'ara axis (Figures 8–10), may reflect mid-Paleozoic epeirogenetic movements associated with "Caledonian" equivalent orogenesis elsewhere. Similarly, regional uplift and non- deposition or erosion during or in post-Carboniferous time, which resulted in attenuation or absence of Carboniferous deposits and even in places in deep removal of older sections down to the Precambrian (Murris, 1980), may reflect Late Paleozoic reaction to a "Hercynian" orogeny at the leading edge of the Gondwanan margin. At that time, "southward" directed subduction of the paleo-Tethys under the Iran and ? Turkey block, seemingly left limited indications of orogenesis in the Oman Mountains (Lovelock et al., 1981) and in central and western? Iran (Berberian and King, 1981). This subduction transferred the Gondwanan plate edge from a passive to an active margin. Sengor (1990) narrowed this deformation episode down to the period Carboniferous to Early Permian and identified the Sanandaj/Sirjan block of Iran as an ensialic magmatic arc—one of a number that discontinuously bordered the northern margin of Gondwana.

By the Late Permian, an extensional phase with crustal stretching and rifting occurred that led to Triassic separation of the central Iranian plate (and other continental fragments) from the rest of Arabian Gondwana along the Zagros line and to opening of the neo-Tethys ocean. The development of an east–west-trending positive area in northern Oman during the Late Carboniferous–Early Permian, which aligns with the Central Arabian Arch further west (Figure 1), may be linked with thermal doming or isostatic uplift prior to rifting and magmatism in the Permian (Blendinger et al., 1990).

Publisher's Note

Credit to the original sources of information and holders of copyright for many of the figures in this volume were erroneously omitted in publication. We profoundly regret any inconvenience this error may cause previous publishers of the figures or the readers of this book.

We also wish to acknowledge the inadvertent omission of the names Cyprus, Lebanon, and Israel from the map included as Enclosure 1.

Please accept our sincere apologies for these errors, and keep this erratum insert with your copy of the book for future reference.

Following are the original sources and copyright holders, who deserve due credit for their contributions to the subject of this volume:

PAGE	FIGURE	CAPTION
8	2	Early Paleozoic plate tectonic reconstructions: Arabia and surroundings.
CREDITS:		1) Reproduced by permission of the University of Chicago, © 1979 by the University of Chicago. 2) Reproduced by permission of the Geological Society from *Geological Society Memoir 12.*

PAGE	FIGURE	CAPTION
9	3	Later Paleozoic plate tectonic reconstructions: Arabia and surroundings.
CREDITS:		1) Reproduced by permission of the University of Chicago, © 1979 by the University of Chicago. 2) Reproduced by permission of the Geological Society from *Geological Society Memoir 12.*

PAGE	FIGURE	CAPTION
10	4	Generalized paleolatitude curve for Oman.
CREDITS:		Prepared by M.W. Hughes Clark, 1984; based on plate tectonic data from Shell International and various published sources. Revised Curve (•—•) (Muscat, 1989) courtesy of Shell International Petroleum, EXP/31.

PAGE	FIGURE	CAPTION
11	5	Infracambrian/Paleozoic formations correlation: southeast Turkey/Oman.
CREDITS:		Adapted from: 1) Beydoun, 1988; Berberian and King, 1981; 2) Lababidi and Hamdan, 1985; 3) Dunnington et al, 1959; 4) Berberian and King, 1981; Stocklin, 1986; 5) Hughes Clarke, 1988; Beydoun, 1988, 1989b. Reproduced from Beydoun (1989b) with permission of OAPEC.

PAGE	FIGURE	CAPTION
12	6	Infracambrian/Paleozoic formations correlation: Jordan—Yemen and Dhofar.
CREDITS:		Adapted from: 1) Beydoun, 1988, 1989b; 2) Powers, et al., 1966; Al-Laboun, 1986; Sharief & Moshirf, 1989; Delfour, 1970; Nebert et al., 1974; Beydoun, 1988, 1989b; Vaslet, 1989; 3) Beydoun, 1988, 1989b. Updated from Beydoun (1989b). Reproduced by permission of OAPEC.

PAGE	FIGURE	CAPTION
13	7	Paleogeography of the Late Precambrian/Early Cambrian sediments of Gondwana.
CREDITS:		Adapted from Gorin et al. (1982) with polar direction modified to that by Scotese et al. (1979) and Scotese & McKerrow (1990).

PAGE	FIGURE	CAPTION
14	8	Arabian plate; approximate thickness of sedimentary cover.
CREDITS:		Modified from Peterson and Wilson (1986). Reproduced from Beydoun (1989b) with permission of OAPEC.

PAGE	FIGURE	CAPTION
15	9	Arabian Plate; approximate thickness of Paleozoic sediments.
CREDITS:		Modified from Peterson and Wilson (1986). Reproduced from Beydoun (1989b) with permission of OAPEC.

PAGE	FIGURE	CAPTION
17	10	Arabian plate; Late Ordovician-Early Silurian probable distribution of facies.
CREDITS:		Modified from Wolfart 1967 (1) & 1981 (2). (Full citations in List of References). 1) By permission of Erdöl und Kohle Erdgas Petrochemie; 2) *In* C.H. Holland, editor, © 1981 by John Wiley & Sons, Ltd.; reproduced by permission of John Wiley & Sons, Ltd.

PAGE	FIGURE	CAPTION
20	11	Mesozoic plate tectonic reconstructions: Arabia and surroundings.
CREDITS:		1) Reproduced by permission of the Geological Society from *Geological Society Special Publication 37*; 2) With permission of Elsevier Science Publishers and the authors; 3) With permission of Cambridge University Press.

PAGE	FIGURE	CAPTION
21	12	Mesozoic formations correlation: southwest Iran—southeast Turkey. For legend see Figure 5.
CREDITS:		Adapted from: 1) James & Wynd, 1965; Setudehnia, 1972; 2) Dunnington et al., 1959; 3) Rigo de Righi & Cortesini, 1964; Open File Reports. Reproduced from Beydoun (1988) by permission of Scientific Press.

PAGE	FIGURE	CAPTION
22	13	Mesozoic formations correlation: offshore Iran/UAE—western Iraq. For legend see Figure 5.
CREDITS:		Adapted from: 1) Mina et al., 1967; 2) Sugden, et al., 1975; 3) Powers, 1968; 4) Dunnington et al., 1959. Reproduced from Beydoun (1988) by permission of Scientific Press.

PAGE	FIGURE	CAPTION
24	14A	Cenozoic formations correlation: A. offshore Iran/UAE—western Iraq.
CREDITS:		Adapted from: 1) Mina et al., 1967; 2) Sugden et al., 1975; 3) Powers et al., 1966; 4) Dunnington et al., 1959. Reproduced from Beydoun (1988) with permission of Scientific Press.
24	14B	Cenozoic formations correlation: B. southwest Iran—southeast Turkey.
CREDITS:		Adapted from: 1) James and Wynd, 1965; Setudehnia, 1972; 2) Dunnington et al., 1959; 3) Rigo de Righi & Cortesini, 1964; Open File reports. Reproduced from Beydoun (1988) with permission of Scientific Press.

PAGE	FIGURE	CAPTION
26	15	Cenozoic plate tectonic reconstructions: Arabia and surroundings.
CREDITS:		1) Reproduced with permission of Elsevier Science Publishers and the authors; 2) Reproduced with permission of Cambridge University Press.

PAGE	FIGURE	CAPTION
28	17	Mesozoic formations correlation: Oman—Yemen. For legend see Figure 5.
CREDITS:		Adapted from: 1) Morton, 1959; Tschopp, 1967; Glennie et al. 1974; Glennie, 1977; Beydoun & Greenwood, 1968; Beydoun 1982, 1988; Hughes Clarke, 1988; 2) Beydoun & Greenwood, 1968; Beydoun, 1964, 1982; 3) Geukens, 1960, 1966; Lamare, 1936. Updated from Beydoun (1988); reproduced by permission of Scientific Press.

PAGE	FIGURE	CAPTION
32	18	Cenozoic formations correlation: Oman—Yemen. For legend see Figure 5.
CREDITS:		Adapted from: 1) Morton, 1959; Tschopp, 1967; Glennie, 1977; Hughes Clarke, 1988; Beydoun & Greenwood, 1968; Montenat et al., 1977; Hawkins et al., 1981; Roger et al., 1989; 2) Beydoun and Greenwood 1968; Beydoun 1964, 1982; 3) Geukens, 1960, 1966; Gillman, 1968. Updated from Beydoun (1988), reproduced by permission of Scientific Press.

PAGE	FIGURE	CAPTION
36	19	Cenozoic plate-tectonic reconstructions: north Arabian margin and Turkey.
CREDITS:		Simplified from Sengör & Yilmaz, 1981. Reproduced with permission of Elsevier Science Publishers and the authors. (Full citations in Reference List.)

PAGE	FIGURE	CAPTION
37	20	Mesozoic formations correlation: northeast Syria—Jordan. For legend see Figure 5.

CREDITS: Adapted from: 1) Daniel, 1963; Weber, 1963; Wetzel, 1974; Bach-Imam & Sigal, 1985; Walley, 1983; 2) Daniel, 1963; Bender, 1974; Heimbach, 1976. Updated from Beydoun (1988); reproduced by permission of Scientific Press.

PAGE	FIGURE	CAPTION
38	21	Cenozoic formations correlation: northeast Syria—Jordan. For legend see Figure 5.

CREDITS: Adapted from: 1) Daniel, 1963; 2) Daniel, 1963; Bender, 1974; Heimbach, 1976. Reproduced from Beydoun (1988) by permission of Scientific Press.

PAGE	FIGURE	CAPTION
42	22	Phanerozoic formations correlation: Gulf of Aden basin.

CREDITS: Adapted from: Beydoun, 1964, 1970, 1982, 1988, 1989a; Beydoun and Greenwood, 1968; Azzaroli, 1957, 1968; Azzaroli and Fois, 1964; S.O.E.C. 1954; Barnes, 1976. Reproduced from Beydoun (1989a) by permission of the *Journal of Petroleum Geology.*

PAGE	FIGURE	CAPTION
43	23	Phanerozoic formations correlation: Red Sea basin and hinterland. For legend see Figure 22.

CREDITS: Adapted from: Gillman et al.,1966; Mulder et al., 1975; USGS/ARAMCO, 1963; Gillman, 1968; Ahmed, 1972; Barakat and Miller, 1984; Savoyat et al., 1989; Bunter and Abdel Magid, 1989. Reproduced from Beydoun (1989a) by permission of the *Journal of Petroleum Geology.*

PAGE	FIGURE	CAPTION
44	24	Red Sea—Gulf of Aden regional map.

CREDITS: Modified from Beydoun (1989a) and reproduced by permission of the *Journal of Petroleum Geology.*

PAGE	FIGURE	CAPTION
51	25	Conceptual sketch map of the "Arabian Promontory," an eastern spur of Africa.

CORRECTION: Conceptual sketch map of the "Arabian Promontory," an eastern spur of Africa, **from the Jurassic to the Eocene**.

Chapter 4

Post-Paleozoic Development: Formation of the Arabian Northeast Passive Margin

The northeast Arabian plate margin initially formed as a result of the Late Permian extension and rifting. However, in the Triassic, this led to the breakup, subsidence and separation of former fragments of marginal Gondwana (represented by central Iran and other blocks, such as the Sanandaj/ Sirjan magmatic arc, northwest Iran and part of Turkey) from the Arabian margin as part of the Cimmerian continent. The initial opening of the neo-Tethys ocean can perhaps be considered a back-arc basin (Sengor, 1990). Although compressed by Tertiary collision, this margin now extends from northwest Iraq, east of the north-trending Ha'il-Ga'ara (-Khleisia) arch of Paleozoic or older origin (Figure 1), to Oman.

An extraordinarily wide, shallow, marine (epeiric) shelf developed on this Mesozoic passive margin. The vast areal extent of this shallow sea (no comparable modern analog on this scale exists), engendered a remarkable uniformity of deposits and the great lateral extent of potential source, reservoir, and seal lithofacies. The southeast (Oman-Somali) margin and the north and northwest (Levant) margins evolved either simultaneously with the northeast margin or very shortly after. Present plate reassemblies require that the southeast margin formed when the Afghan and perhaps part of central Iran block separated from between Arabia and India along transform fractures, as the breakup of western and eastern Gondwana proceeded. The north and northwest margins formed by propagation of the neo-Tethys spreading center northwestwards, and by drift of the main Turkish blocks from the Levant northwards (Figure 11).

Arid and semi-arid conditions characterized the Triassic, and red beds, shallow-water carbonates and interbedded evaporites and shales were the principal deposits. However, deeper-water carbonates and terrigenous clastics characterized the outward shelf/ slope on the one hand and the landward side of the shelf on the other (Figures 12 and 13). Stable shelf conditions typified the Jurassic as well as much of the Cretaceous, with shallow-water carbonates accumulating on the shelf and inner platform. These carbonates were interspersed with areas of intra-shelf euxinic deposits and widespread terminal Jurassic cyclic sabkha-type interbedded evaporites. Concurrently, turbidites accumulated on the continental rise and oceanic cherts and radiolarites on the widening neo-Tethys ocean floor. By the end of the Jurassic the Kirsehir (Turkish) fragment had drifted northwards and the oceanic domain had propagated to extend the neo-Tethys along the northern margin (Dercourt et al., 1986) (Figure 11B).

Latest Jurassic (to earliest Cretaceous, locally) movements along the two extreme ends of the northeast margin resulted in differential block uplifts, often on a north-south trend, and at times led to deep erosion of the elevated blocks, but the greater length of the margin in between continued to subside. During the Early Cretaceous, when Central Arabia lay athwart the humid equatorial belt, a wider spread of continental clastics from hinterland uplifts and the Arabian shield extended in tongue-shaped deposits into the inner areas of the northeast shelf. These deposits interfingered with more marine clastics and laterally graded out into marine carbonates. However, by the Aptian, a partial return to transgressive shallow marine conditions locally led to widespread neritic carbonate deposition.

From the Albian to the later Cretaceous was a period of predominantly marine carbonate sedimentation. However, the overall transgressive rise was far from regular and was interspersed with local and more regional regressive intervals leaving behind coarser clastics and fluvio-deltaic sands or shallow carbonate reef buildups. Cenomanian and lower Turonian rudistid reefs covered extensive linear trends. Sea level oscillations were punctuated by phases of open-marine conditions with upwellings and anoxia. During these latter events several levels of extensively distributed organic-rich sediments were laid down juxtaposed with good carbonate and sandstone reservoir rocks (Figures 12 and 13). The predominance of carbonates during most of the Mesozoic reflects the paleolatitudinal position of Arabia within tropical to equatorial climatic belts (Figures 11 and 4).

Tectonic deformation of this long and very wide margin began in association with the onset of the emplacement of ophiolites on the Arabian shelf during the late Turonian. Ricou (1971) referred to the ophiolite occurrences extending from the northwest Arabian margin along the edge of Arabia to the southeast margin in Oman as the peri-Arabian ophiolitic crescent. He stressed the importance of thrusting that resulted from the major tectonic events occurring during the

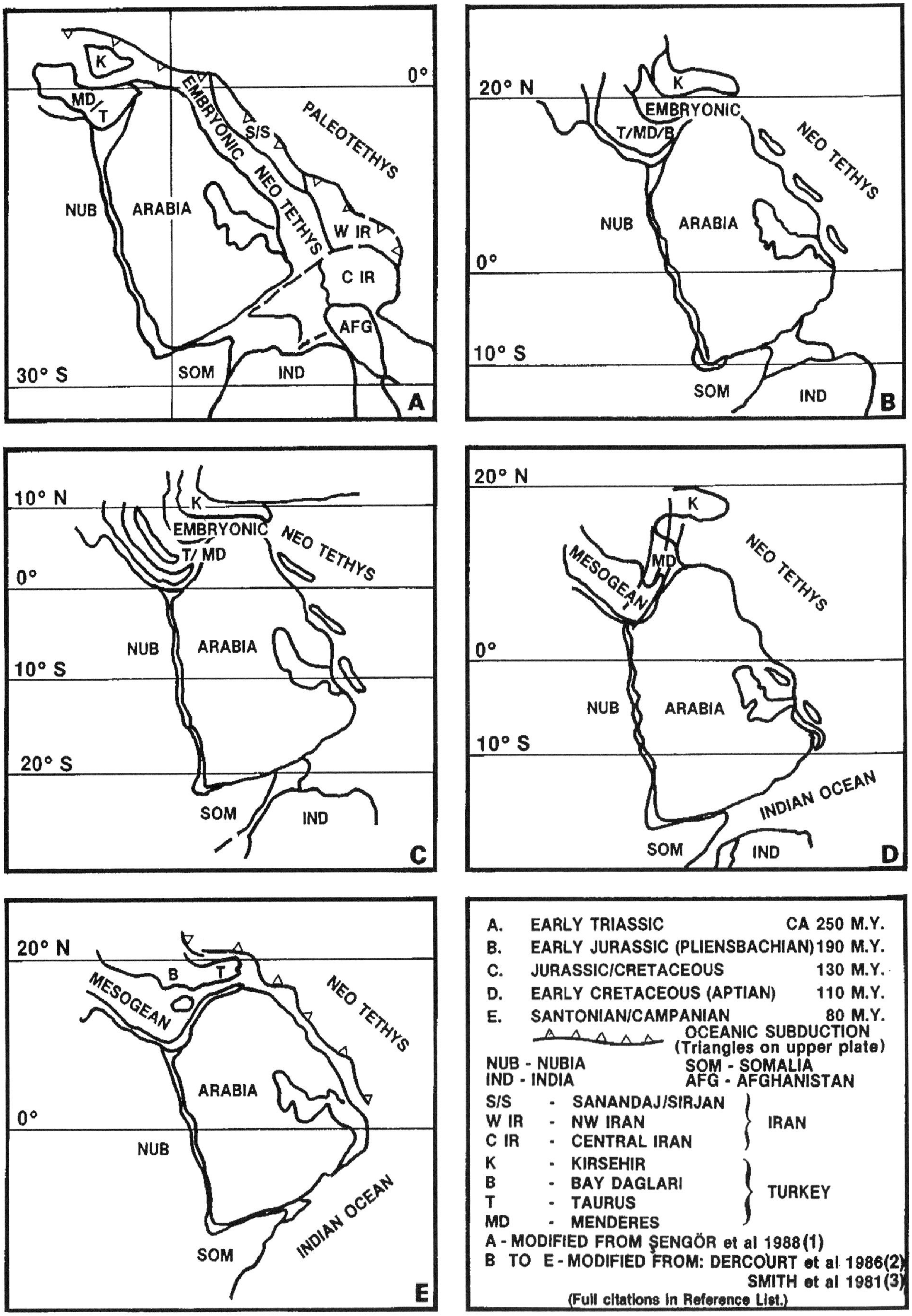

Figure 11—Mesozoic plate reconstructions: Arabia and surroundings.

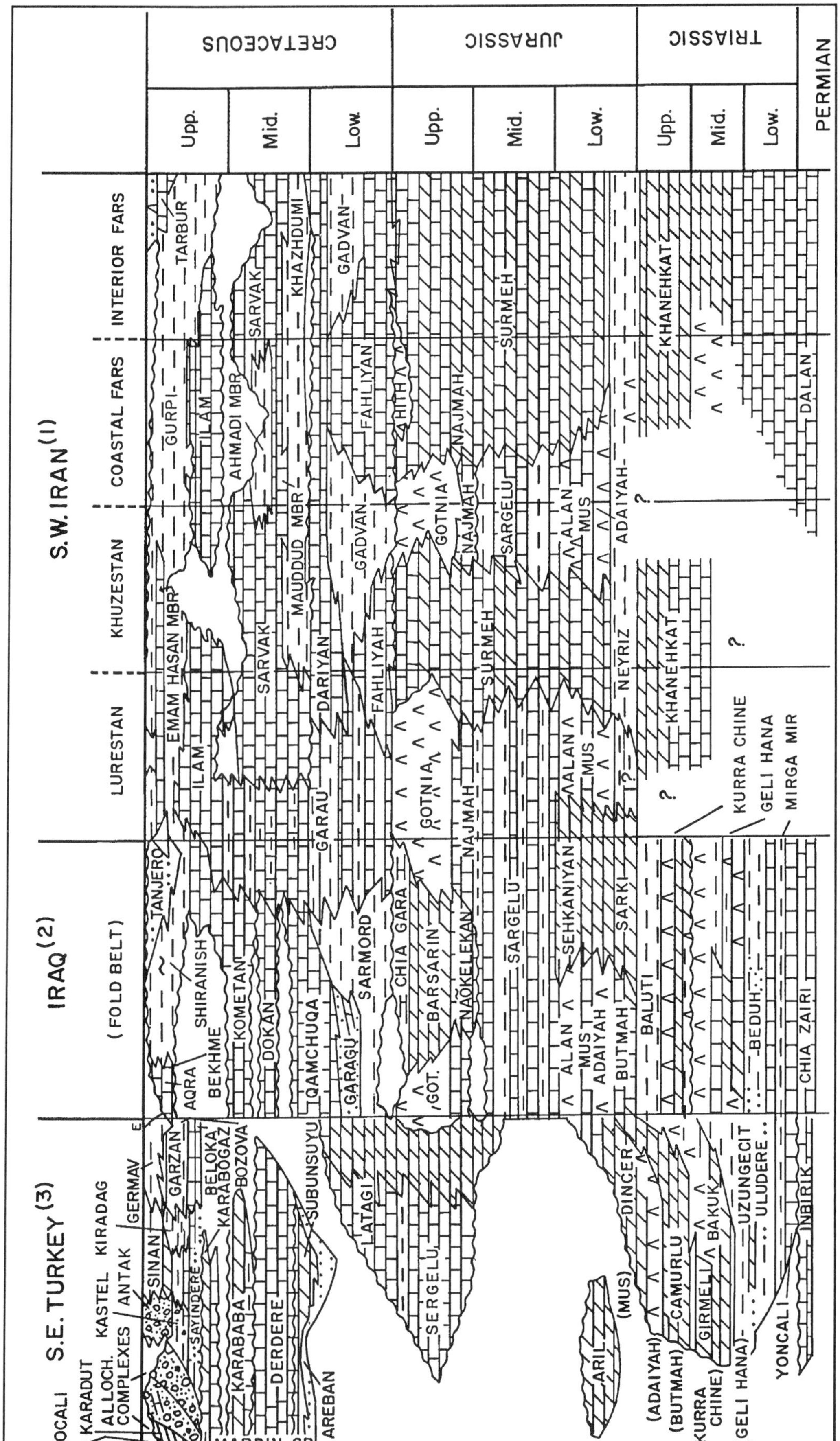

Figure 12—Mesozoic formations correlation: southwest Iran-southeast Turkey. For legend see Figure 5.

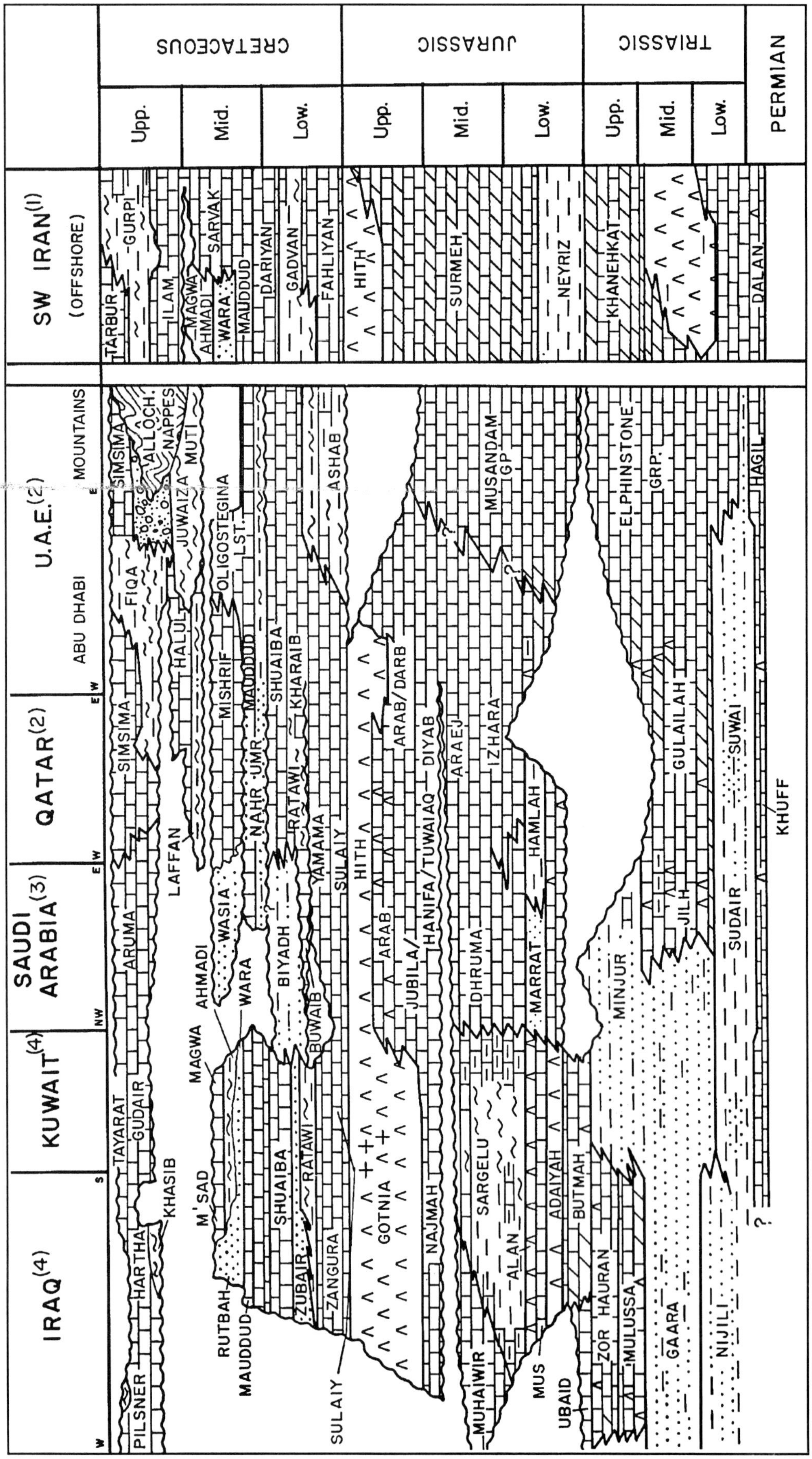

Figure 13—Mesozoic formations correlation: offshore Iran/UAE–western Iraq. For legend see Figure 5.

latest Cretaceous; however, he did not go deeply into the reasons behind the causes (Figure 1). Takin (1972), following earlier models developed for Cyprus (e.g. Gass and Masson-Smith, 1963), was the first to introduce the concepts of plate tectonics in discussion of the northeast margin. He postulated the development of a subduction zone along part of the southern edge of Eurasia, which subducted the oceanic crust between Arabia and Eurasia prior to the Maastrichtian. This subduction was terminated with the emplacement of part of that crust as ophiolites on the northeast edge of the Arabian plate. Glennie et al. (1973, 1974), as a result of detailed field work in Oman, inferred the development of three adjoining sedimentary provinces on both the shelf and slope and off the northeast margin of Afro-Arabia, all with depositional histories from Late Permian time. They postulated shallow marine deposits on the shelf, turbidites on the slope and rise, and abyssal deposits in the ocean basin. During the Late Cretaceous, depositional units from the last two provinces, together with part of the crust from the oceanic ridge, were tectonically emplaced as a series of imbricate nappes onto the deposits of the Arabian shelf as a consequence of the closing of the bordering ocean.

Since that time a considerable amount of literature has become available proposing a number of differing models (several, up until 1985, were reviewed by Beydoun [1988] and details of these will not be dealt with here). This literature represents a wealth of data on the Afro-Arabian late Mesozoic Tethyan margin in general and the northeast Arabian margin in particular. Those highlighting the Late Cretaceous deformation and obduction events along the northeast margin are summarized here.

The model proposed by Smith and Woodcock (1976) calls for southward subduction below the Arabian plate by dehydration of a sinking mantle slab. Haynes and McQuillan (1974) invoked rupture and upward thrusting of slices of oceanic crust by a process of uncoupling, with the cracked and uncoupled parts of the northward subducting plate (under the stable Persian continent) being underthrust by the pushing plate still attached to the Arabian continent. The uncoupled slices were thus being aided upward and rising by isostatic response to finally be manifest as a belt of ultramafics in the trench zone on the Persian block. They argued that it is Arabian plate collisional movements in the Neogene that resulted in thrusting and overfolding in the imbricated zone and thus, there is no need to invoke Late Cretaceous collision. Conversely, Welland and Mitchell (1977) called for a northward-dipping subduction zone under the Iranian region, deformation of that margin, and incorporation of basinal and margin facies into a lower trench-slope accretionary prism that later incorporated ophiolite sheets on arrival of the ridge at the trench. The whole tectono-sedimentary package then was emplaced on the Arabian platform by collision of the edge of Arabia with Iran in the Late Cretaceous. Hutchinson et al. (1981) favored an alternative model that suggests that the (Semail) ophiolite represents backarc crust emplaced on the Arabian margin during continental underthrusting in the Late Cretaceous. The present Gulf of Oman represents the northeastern continuation of the oceanic crust outcropping in the Semail ophiolite. Gealey (1977) attributed the obduction of the Oman ophiolite and evolution of the Oman Mountains to collision between a passive Arabian margin and an island arc; the ophiolite emplacement was due to the underthrusting of the Arabian margin under the forearc limb. Later the slab was uplifted because of shortening and isostatic adjustment of the underthrusting continental crust. Alavi (1980) envisaged the Arabian block moving to the northeast as the leading edge of a continent trailing behind a subducting oceanic crust going down a trench in front of the Iranian continent. By the Late Cretaceous, progressive loss of oceanic crust along the subduction zone led to early phases of collision of the Iranian block and Arabian continent. Southwestward obduction of ocean floor slices along high-angle drag faults was followed by synthetic low-angle thrust faulting. This faulting accompanied and followed folding and uplift of oceanic sediments, which moved sheets of allochthonous material onto the Arabian platform. An Andean type magmatic arc developed along the borders of the Iranian block.

Coleman (1981), using data constraints on the age of the Semail ophiolite that were not taken into account in some of the earlier models, showed that the Oman ophiolite was part of a spreading center within a restricted Tethys and unrelated to continental margins and island arcs invoked by some of those models. His model called for Cretaceous spreading north of Arabia during which, nevertheless, the Afro-Arabian plate moved northwards. Eurasia and Africa rotated in opposite directions and initiated closure of the Tethys during the Late Cretaceous. In the early stages, the Arabian margin was downwarped, which, together with compression from closure, initiated obduction of ocean crust along existing transform faults with still-hot crust being detached along oblique, northeast-dipping thrust faults and emplaced by gravity sliding across the Arabian margin. Subduction and large-scale island arc vulcanism began after that; reworked laterites (evidencing emergence) rest on the eroded ophiolite surface and are in turn overlain by shallow-marine Maastrichtian deposits.

Since my review (Beydoun, 1988), substantial additional documentation concerning the northeast margin has appeared in special conference proceedings or special issues of journals and has principally related to the Oman portion of the northeast margin but has included data from Iran. Notable among these is the memoir by Lippard et al. (1986) synthesizing several years of field work by the Open University and other teams. Lippard et al. (1986) refined the results of Glennie et al. (1973, 1974) particularly with regard to questions of unrealisticly high previous displacement rates of thrusting/obduction and rapid changes between overall consumption and spreading. They also made fur-

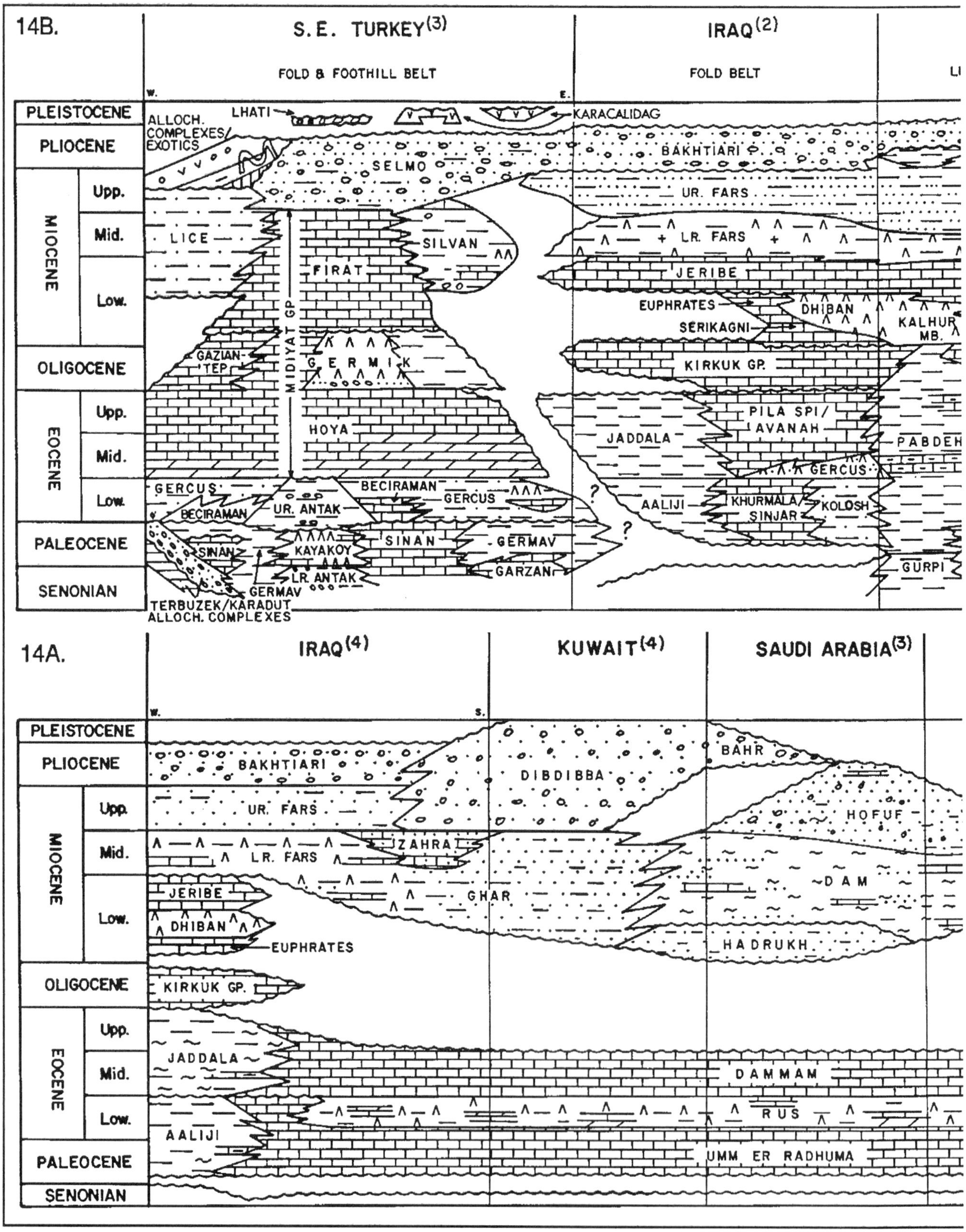

Figure 14—Cenozoic formations correlation: A. offshore Iran/UAE-western Iraq B. southwest Iran–southeast Turkey.

ther refinements to the tectonostratigraphy and sequential order of the deformational tectonic events.

The emplacement episodes occurred from the late Turonian to the early Maastrichtian, with the final

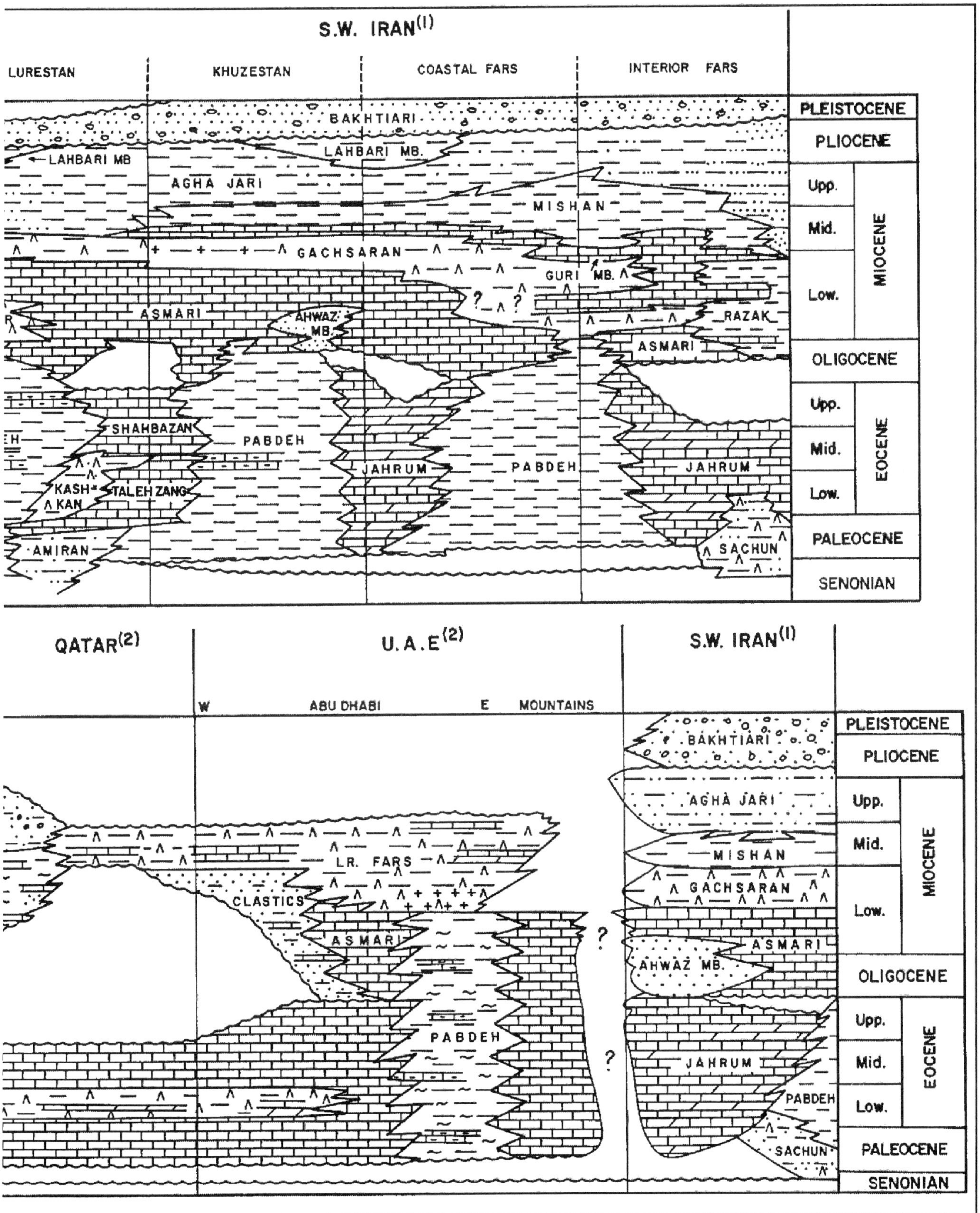

ophiolite obduction occurring in the last 10 Ma time span (Kazmin et al., 1986) (Figure 11). Upper Maastrichtian deposits lie unconformably on the obducted greenstones with late Maastrichtian and Paleocene

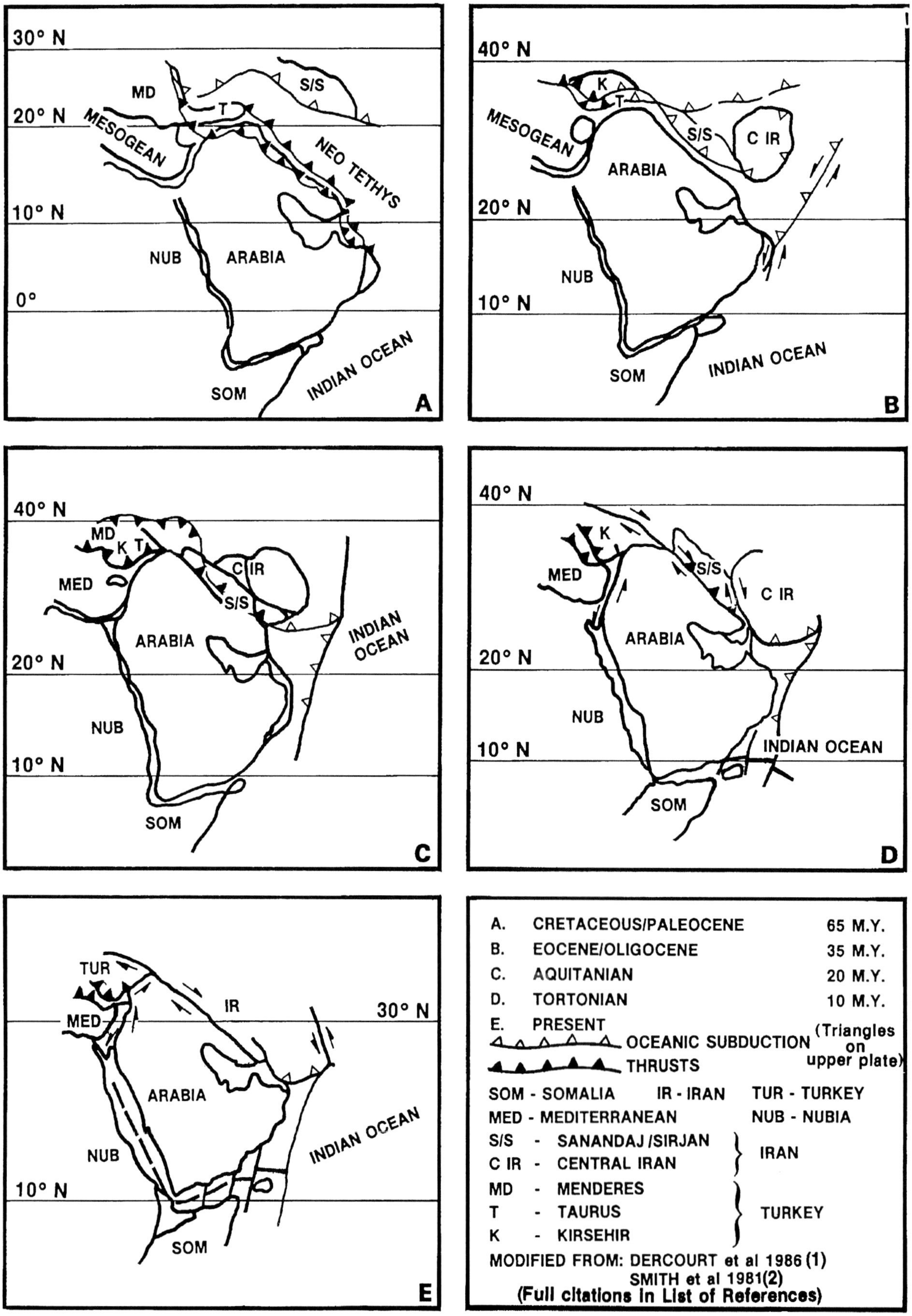

Figure 15—Cenozoic plate tectonic reconstructions: Arabia and surroundings.

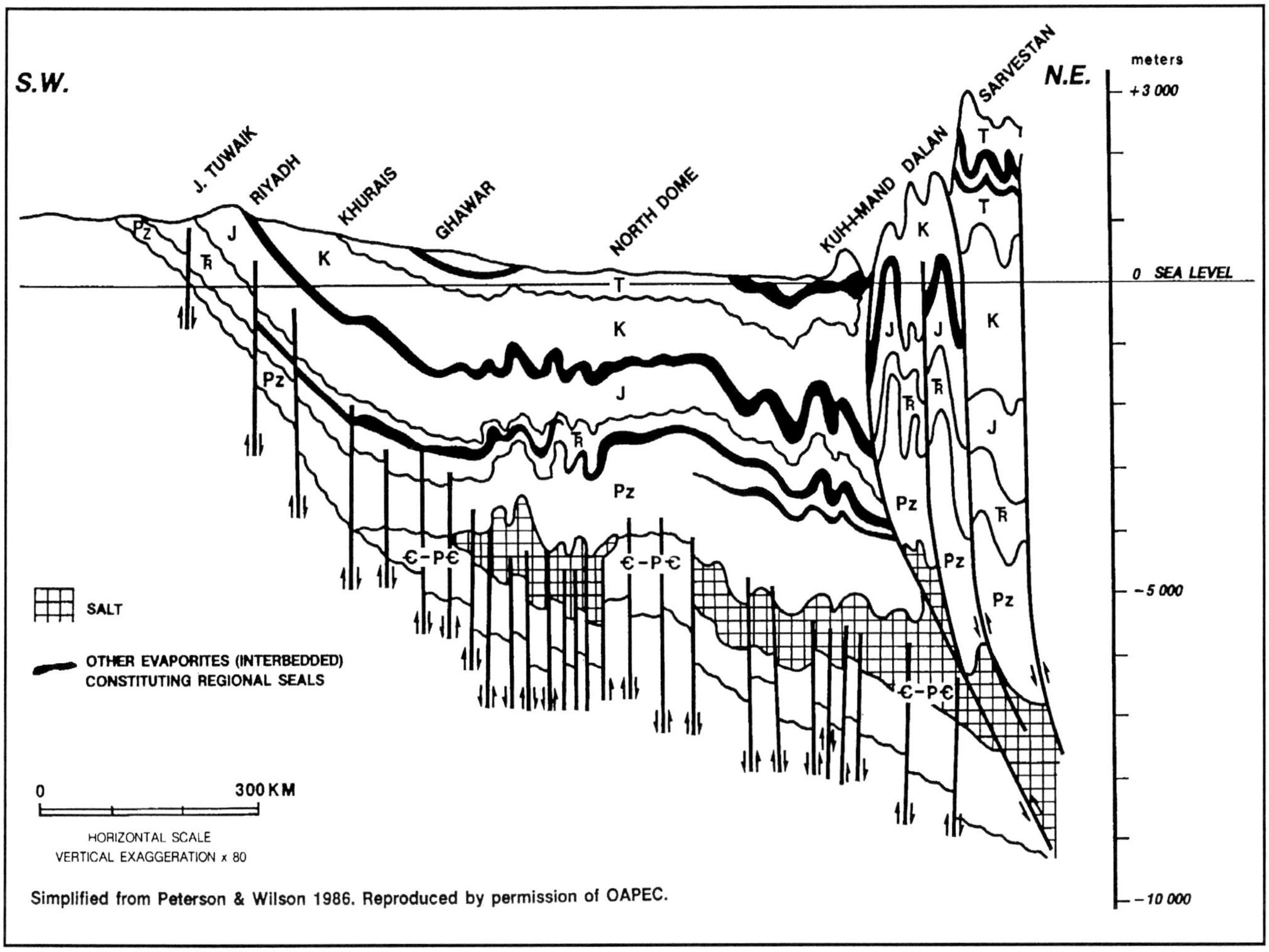

Figure 16—Southwest-northeast structural-stratigraphic cross section, northeast Arabian plate margin.

transgressions inundating the deformed margin (Figures 12 and 13). Some lateral shear of plate edge fragments from the northeast margin to the north margin probably occurred (A.M.C. Sengor, personal communication, 1990).

Sengor et al. (1988) preferred to explain this episode of thrusting and obduction as being the result of the formation in Early to middle Cretaceous time of north-dipping subduction zones in Turkey, Iran and Oman (easterly dipping in the Oman sector according to Lippard et al., 1986 and Glennie et al., 1990). They suggested that these events were short-lived because they originated too close to the north-facing Gondwanan passive continental margin and collided with it, leading to widespread ophiolite obduction. Stoneley (1990a), on the other hand, proposed a different mechanism (at least for the Iranian–Oman sectors) linked to extensional faulting parallel to the margin, which culminated in the Late Cretaceous with the expulsion of mantle material to the surface. This engulfed the shelf edge before gliding southwestward over complementary intra-shelf basins, and it occurred progressively later toward the northwest. He interpreted these events as an unsuccessful attempt to develop a spreading axis along the continental margin and suggested that the Oman and Iranian sectors of the Arabian margin were offset 300 km by a transform fault. That would explain the differences between the two regions as being due to attempted spreading in different tectonic environments.

The Oman sector obduction scenario of Glennie et al. (1990) envisaged the buildup of an accretionary wedge of Hawasina (oceanic sediments and Colored Melange) as Tethyan oceanic crust was consumed in the easterly dipping subduction zone until the Late Cretaceous. At that time, the buoyant Arabian continental margin crust appears to have choked the subduction process. Continuing convergence caused acceleration or creation of another subduction trench between, and extending south and east of, the Iranian microcontinents (envisaged as already being close across the diminishing neo-Tethys) from which the present Makran subduction is a remnant. The resulting crustal relaxation on the Arabian margin subduction zone permitted the uplift of the "half-buried" Oman continental margin, causing the Hawasina wedge and

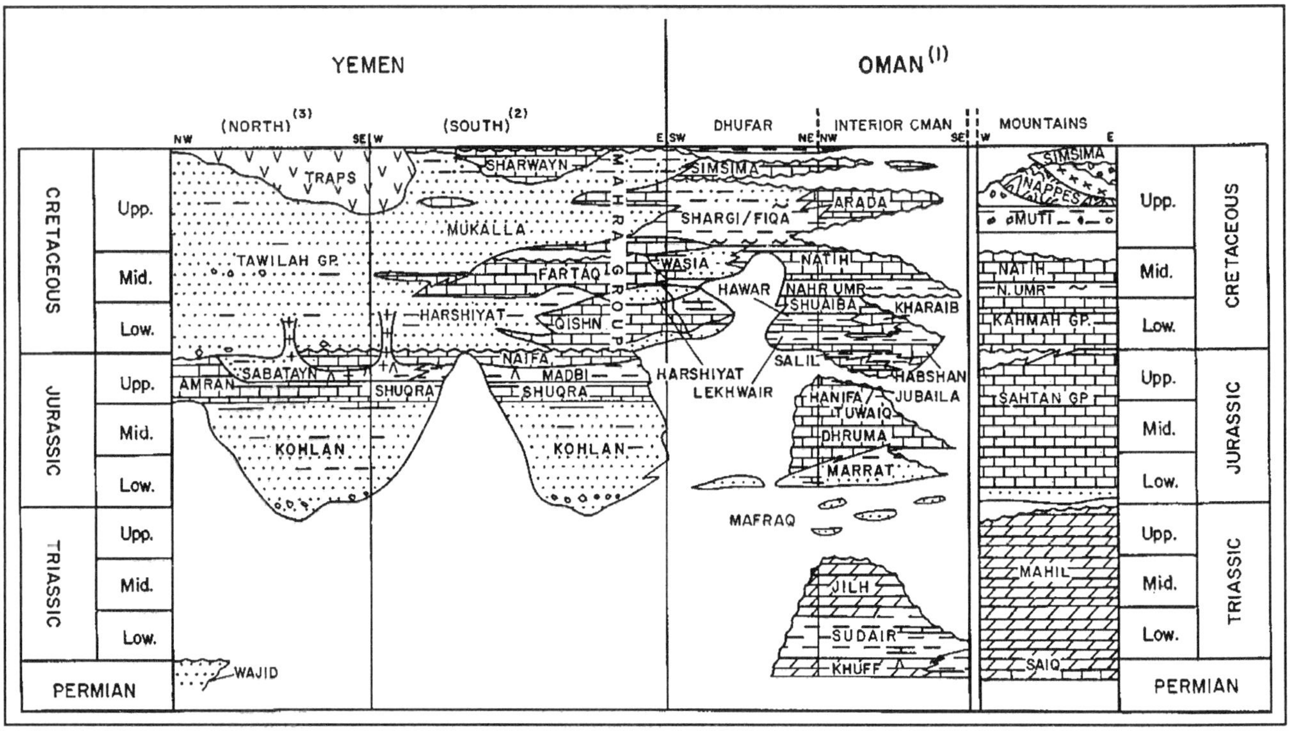

Figure 17—Mesozoic formations correlation: Oman–Yemen. For legend see Figure 5.

former hanging wall of the Semail ophiolite to spread gravitationally farther onto the Arabian shelf.

Kazmin et al. (1986) recognized a number of structural elements from ocean to shelf traverses along the northeast margin: (a) a flysch trough (? remnant of Tethys floor); (b) an uplifted, periodically emerging zone (e.g. the Fars ridge in the Zagros); (c) a border furrow with somewhat deeper water sedimentation (such as can be seen in northern Iraq); and (d) the shallow and very wide Arabian shelf (underlain by stretched continental crust, Hempton, 1987).

As the foregoing review shows, the extent and type of subduction and the type and degree of collision that gave rise to the Late Cretaceous deformation and ophiolite obduction all along the northeastern Arabian margin and the adjacent northern and southeastern margins are subject to debate.

Whatever the details of the various models for the Late Cretaceous obduction events along not only what had become the leading edge of the northeast margin but also to a lesser extent along the leading edge of the north margin (which is discussed in following sections), the consequences of these events for the hydrocarbon habitat were important and can be summarized as follows:

1. The stable shelf was disturbed after long quiescence.

2. The outer margin was overrun by nappes and underlying sediments were "overcooked."

3. On the inner shelf, older structures were rejuvenated, particularly those associated with salt; this was a time of new trap formation or increase in earlier trap amplitudes.

4. Cycles of deeper-water facies, including shales (many of which are bentonitic, perhaps from island-arc tuffs) were spread across the shelf partially forming good hydrocarbon seals.

5. A wide regression occurred at the end of the Cretaceous as strain was released and the plate bounced up, and hydrodynamic gradients were activated.

The Cenozoic era opened with a powerful transgression covering the entire shelf, deep into the hinterland, leaving behind neritic carbonates and/or pelagic deposits (marls and shales) along the shelf edge, including organic-rich source rocks (Figure 14). Shallower transgressive/regressive carbonate and evaporitic shelf conditions followed in the Eocene, but deeper-water pelagic deposits were laid down in parts of southwest Iran/northeast Iraq and Oman and continued into the Oligocene. This deposition occurred adjacent to the ophiolite front. Although much of the Arabian plate was uplifted in the late Eocene (Arabia was well into a northern hemisphere tropical/temperate position–Figure 15), the northeast/northern margins maintained a narrowing and shoaling seaway well into the Miocene, linking the Mediterranean to the west and the Indian Ocean to the southeast. At times the seaway was cut off from oceanic flushing, which gave rise to evaporites interbedded with clastic red beds as well as shallow-water deposits and reef carbonates. This combination formed good hydrocarbon reservoir and seal units (Figure 14).

With Arabia beginning to move northeastward from Africa in the Miocene (Figure 15), this narrow-margin seaway was closed and the "passive" margin leading edge became a fully collisional margin. Uneven convergence of the Arabian and Eurasian plates resulted in compressional foreland folding and more complex deformation in the rising orogens, as a result of decollement in the deeper evaporitic levels (Figure 16). In the Pliocene-Pleistocene, molasse was shed from the rising Zagros and Taurus ranges as suturing was accompanied by increased thrusting. The motion between Arabia and the Iran/Turkey parts of the Eurasian plate was taken up only in part by this compressional shortening. Additionally, as stress built up from continuing northward drift of the Arabian plate, motion was also taken up by strike-slip dextral motion along the Zagros crush zone and by dextral and sinistral movement on the North and East Anatolian faults.

Total sediment thickness along the entire northeast margin can reach up to 13,000 m at the plate edge (45,000 feet) and is at least 2,500 m on the interior side of the margin nearer the Arabian shield (Figures 8 and 17). Tertiary structural deformation over the greater part of the margin is basement controlled by block faulting away from the Zagros orogen, while compression is assisted by flow or decollement in deep and shallow evaporites near the Zagros suture, with high-angle reverse faulting and thrust stacking becoming increasingly intense toward the suture.

Chapter 5

Post-Paleozoic Development: Formation of the Arabian Southeast Margin

Throughout the Paleozoic, the region now forming the southeast Arabian plate margin was an internal part of the northern Gondwanan (paleo) Tethyan margin, together with other microplates, blocks, or fragments that separated from the Gondwanan margin in the Triassic to partly form the Cimmerian continent. The southeast margin's depositional history was similar to that of the rest of the Arabian margin and the adjoining portions, with tectonic deformation in the late Paleozoic related to a relatively short period of paleo-Tethys subduction under the southern (Gondwanan) margin and doming in Carboniferous–Early Permian time (see Chapter 3, Figures 2, 3, 5, 6).

Present Paleozoic plate reconstructions leave some question as to which continental elements constituted the outer margin of the part of Gondwana adjacent to Arabia. For instance, we are unsure whether the central Iran microcontinent was adjacent to the Afghan block as in Figure 11A or was as shown in Figure 3. By the Late Permian, the entire Arabian Gondwanan margin was under a full extensional regime. By the Triassic this extension led to rifting and separation of the outer fragments from Arabia along the Zagros line. Initially, this separation may have been in response to spreading in a back-arc basin (Sengor, 1990). If the Afghan and central Iran blocks originally lay between Arabia and India, they must have moved away along transforms during the Early Triassic rifting and partial break-up of West and East Gondwana (Figure 11A). This would have led to the development of the original southeast Arabian plate margin as a transform margin.

After subsidence and continued deposition during the early Paleozoic, this margin underwent a first phase of uplift during the late Paleozoic. The result was truncation of lower Paleozoic units toward the southeast beneath the Carboniferous-Permian glacials (M.W. Hughes Clarke, personal communication, 1990). A further phase of uplift of the southeast margin occurred in the Permian-Triassic, with updoming prior to Indo-Arabian rifting, followed by separation of the Afghan/Iran/India blocks (Shackleton and Ries, 1990) that led to the formation of a proto-Owen basin in the Jurassic.

During the Jurassic, the southeast margin continued relatively to be positive. In the terminal Jurassic–Early Cretaceous, further Afro-Indian divergence resulted in India moving away from Somalia and a southeast Arabian passive margin extended from Dhofar in southwest Oman to the Somalia sector and was accompanied by the formation of the northern Somali basin (Figure 11C and 11D).

In the Early Cretaceous, the Masirah trend ophiolites of the southeast margin apparently were uplifted and unconformably covered by limestone. More certainly, they were upthrust and obducted westwards in the Late Cretaceous, perhaps in movements associated with the Semail (Oman Mountains) ophiolite obduction. In this interpretation, deformation along the margin resolved into right lateral strike-slip motion on the Masirah fault and the west-northwest verging Batain fold and thrust belt of the southeast margin (Shackleton and Ries, 1990). Recent (1986) marine seismic work in the Owen basin, however, has led Mountain and Prell (1990) to question the Jurassic-Early Cretaceous age of the crust in the present Owen basin. To reconcile their observations with the established facts concerning the opening of the northwest Indian Ocean, they proposed the following scenario: A proto Owen basin formed in Jurassic time as a result of the separations discussed above. Following the Late Cretaceous separation of India from Madagascar, however, the new spreading ridge extended to contact the southeast Oman margin and the proto Owen basin crust was carried northeastward toward eventual subduction. This resulted in formation of a strike-slip Oman margin adjacent to a Late Cretaceous Owen basin. The ophiolites on the southeast Oman coast are fragments of this crust, emplaced most probably as a result of left-lateral transpressional forces along this plate boundary.

In the Neogene, the Gulf of Aden was opened by the propagation of a new spreading center. Major compressional stresses arising from this reorganization of plate geometries led to the elevation of the Owen ridge and reactivation of Late Cretaceous faults along the Oman margin (Mountain and Prell, 1990). The Masirah trough probably formed then and both strike-slip and horst and graben forming motions occurred all along this southeast margin, culminating in its further uplift. Strike-slip motion along the Masirah fault occurred again late in the Neogene.

The Infracambrian through Paleozoic succession of the Arabian plate, including the southeast margin, has already been correlated and discussed in earlier sections (Figures 5 and 6). Nevertheless, the Infracambri-

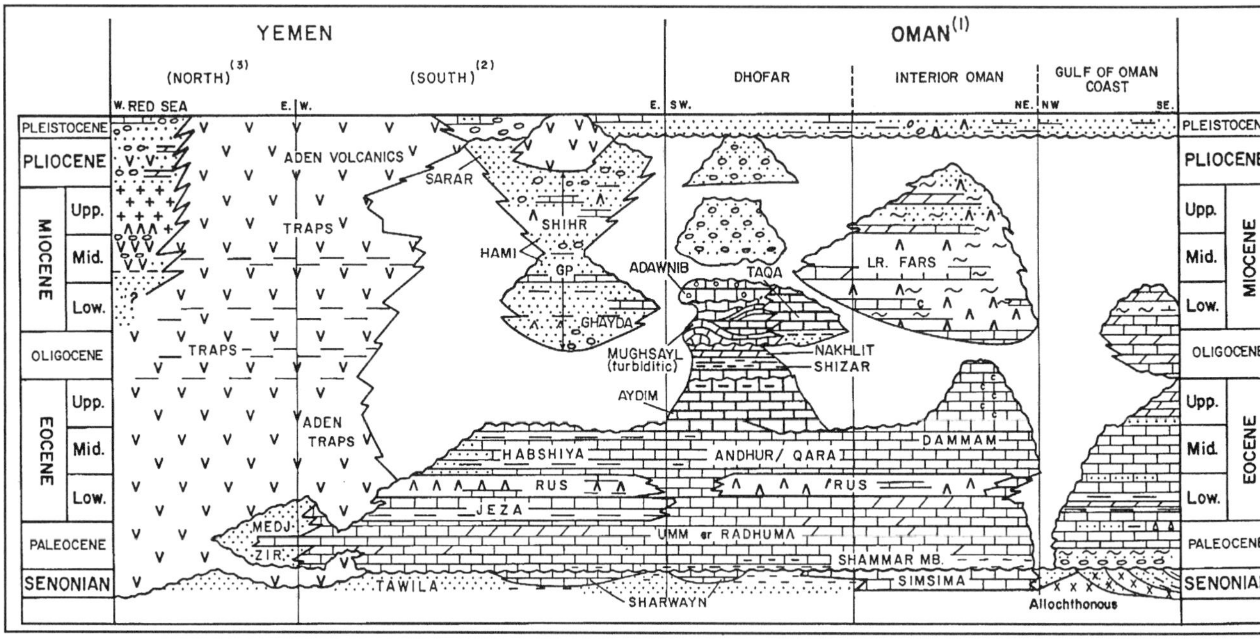

Figure 18—Cenozoic formations correlation: Oman–Yemen. For legend see Figure 5.

an/Lower Cambrian Huqf group succession of the southeast margin merits additional remarks in view of its petroleum source, reservoir and seal potential and in order to understand the link between the earlier surface and the latest subsurface data. Mattes and Conway Morris (1990) have addressed this problem particularly because the Ara evaporitic formation is essentially absent at the surface. They differentiate between the outcrop sequence and subsurface scheme. In the latter, the Ara consists of a series of dolomites and evaporites, with characters that suggest an age of 550-570 Ma as well as a thick pre-salt carbonate sequence, which may be wholly or partially equivalent to the upper part of the Huqf outcrop sequence. A subsurface pre-Ara section is separated from the Ara, as defined above, by an angular unconformity. Thus, in this new subsurface definition of the Ara, shallow-water Ara carbonates provide reservoir rocks of widely varying quality and deeper-water Ara carbonates and shales provide the source rocks that charged not only the shallow-water Ara deposits, but also the overlying Paleozoic clastics. Ara salt forms the top, base, and lateral seal for intra-Ara hydrocarbon pools.

A major hiatus from the Early Silurian to the Late Carboniferous is present in the Paleozoic succession with the exception of the limited occurrence of the probably lacustrine Devonian Misfar formation. This hiatus reflects episodes of intermittent uplift that affected the region from early Paleozoic time. Outcrops occur along the Huqf-Haushi-Ja'alan axis, which seems to have acted almost continuously as a positive area compared with the adjacent basins (Glennie, 1977). Outcrops also are present in the Murbat area of Dhofar in southwest Oman, but otherwise most characteristics of the Paleozoic succession have been defined from subsurface well information (Hughes Clarke, 1988). The age of the Murbat sandstone formation of Dhofar, which contains basal glaciogenic deposits, is controversial. It has been lithologically correlated with the Abu Mahara formation of the Huqf group in the absence of any palynomorphs (Gorin et al., 1982). More recently, the Murbat has been correlated with the Permian-Carboniferous Haushi clastics, based on apparent Devonian radiometric ages in the underlying crystalline basement and on the presence of glaciogenic deposits (Qidwai, 1988). In my opinion, the latter correlation is the more probable particularly because during a 1954 field visit the sandstone unit immediately above yielded a single lamellibranch cast (courtesy of Dhofar Cities Service Oil Co.). This finding was recently confirmed by one of the field team at that time (T. E. Stanzel, personal communication, 1988; see also Beydoun, 1988, for further discussion).

In general, the Mesozoic succession of the southeast margin is thin and units frequently pinch out to the southeast compared with the sequence in central and northern Oman and in the Oman Mountains (Hughes Clarke, 1988) (Figure 17). The Late Permian Khuff carbonate formation is either absent or thinly developed in a clastic facies. The Triassic-Jurassic units are also thin, have many breaks, and are mainly clastic facies in the southeast; they become predominantly carbonate toward the north. In Dhofar, the entire pre-Cretaceous sequence above the Murbat formation is missing. The Cretaceous sequence is very incomplete in eastern Dhofar but becomes progressively more complete toward the west. The Gulf of Aden to the west was not yet open and the Arabian and Somali sectors formed the distal part of the Afro-Arabian Gondwanan Tethyan shelf during Jurassic/Cretaceous times. This shelf was inundated principally from the east and northeast (Figure 11).

During the Cenozoic, the southeast margin received extensive shelf carbonate sedimentation, with an interbedded evaporite episode during the Eocene and several shale intervals (Figure 18) that form potentially important source rocks. Generally, the greater part of the coastal area of the southeast margin continued to be submerged well into the early and middle Miocene. Marine or lacustrine and restricted circulation deposits reached appreciable thickness, especially in the foundering offshore troughs (Masira-Kuria Muria-offshore Dhofar; see Platel and Roger, 1989; Roger et al., 1989), where potential organic rich Eocene source rocks (e.g. Andhur formation) may now be buried deep enough to be in the oil generation phase.

Chapter 6

Post-Paleozoic Development: Formation of the Arabian North and Northwest Margins and Adjacent Intra-Plate Basins

The Late Permian to Middle Triassic rifting events that led to the formation of the northeast Arabian margin also gave rise to the north and northwest margins by breakup and separation of former portions of Gondwana. These are now represented by Turkey (sensu latu) and also probably by northwest Iran (Sengor et al., 1984). The neo-Tethyan spreading ridge propagated northwestward in the Early Triassic. Along the edge of the present Libyan-Egyptian shelf, rifting followed by sea-floor spreading and formation of an accretionary plate margin developed in the Late Triassic (Carnian) to the Early Jurassic. This opened up a subordinate southern branch of the neo-Tethys and initiated the separation of Turkey from Africa and the Levant. A transform boundary developed that paralleled the shelf edge of the Levant (Dewey et al., 1973) and gradually linked this subordinate arm with the main southern Tethyan (Zagros ocean) branch. This event had already separated the Bitlis-Poturge fragments and the Anatolid-Taurid platform of the Cimmerian continent from this newly formed north Arabian passive margin, (Figure 11B, not specifically identified but they form part of "K" on that figure). Coincidentally the multi-armed northern branch of the neo-Tethys developed initially as a marginal basin resulting from the Early Jurassic internal disintegration of the Cimmerian continent behind the paleo-Tethys subduction zone (Sengor, 1979; Sengor and Yilmaz, 1981).

The northwest (Levant) transform margin did not develop fully until the Early Jurassic (or Early Cretaceous?; Dercourt et al., 1986) (see Figure 11D) when all the Turkish blocks had separated and moved away from Arabian/African Gondwana. In the southern Levant, separation had already commenced in the Late Triassic as indicated by the presence of Jurassic sedimentary prisms reflecting sharp facies changes along a hinge belt between continental shelf and slope/rise environments (Bein and Gvirtzman, 1977). Opinions differ, but once the two arms of the southern Tethys branch had linked up in either the Early Jurassic, (Sengor et al., 1984) or the Early Cretaceous (Dercourt et al., 1986) (Figure 11D), this margin became a passive margin and Jurassic and Cretaceous carbonate reefal buildups characterized the shelf edge depositional facies, interrupted by a Late Jurassic uplift and some erosion. It continued as the Arabian northwest margin until the start of the Neogene, even after closure of the northern seaway linking the eastern Mediterranean with the Zagros basin. With activation of the Levant (Dead Sea transform) fracture in the early Neogene, the old northwest margin ceased to act as the northwest Arabian plate boundary and the "Dead Sea" transform became the new northwest plate boundary separating the Arabian plate from the Levant-Sinai fragment (Figure 1, inset).

The north margin by contrast was block-faulted and partly uplifted from the end of the Jurassic to the Early Cretaceous and the elevated blocks (horsts, swells) were eroded down to the lower Paleozoic in southeast Turkey and northern Syria. This was probably related to the drift of the remaining Turkish blocks by transform motion from the Levant (Figure 11D). This also affected the Levant margin where similar age block uplift occurred or/and was also perhaps related to regional extensional movements in northeast Africa and elsewhere in the area. At this time, block movements were also taking place at the southern end of the Arabian region, probably related to the break-away of India from Afro-Arabia. Aptian-Albian sediments mark the first marine transgression onto this peneplane on both the north and northwest margins.

The latest Cretaceous was a time of extensive ophiolite obduction along the north Arabian platform as well as on the Anatolid-Taurid platform to the north. Obduction on the northeast Arabian platform margin and on the Bitlis-Poturge continental fragment probably began somewhat earlier (ca. mid-Cretaceous). Much of the burial metamorphism of the Bitlis massif is associated with this event. A south-dipping subduction zone beneath the Bitlis-Poturge fragments (Figure 19A) began consuming the Inner Tauride ocean either synchronously or immediately after the obduction as evidenced by lithologies on the Yuksekova arc. The southernmost of these Late Cretaceous events in the region was the nappe emplacement extending from Cyprus along the entire northern margin, and is in general late Campanian/early Maastrichtian in age. This is the age bracket of the Kastel flysch, which

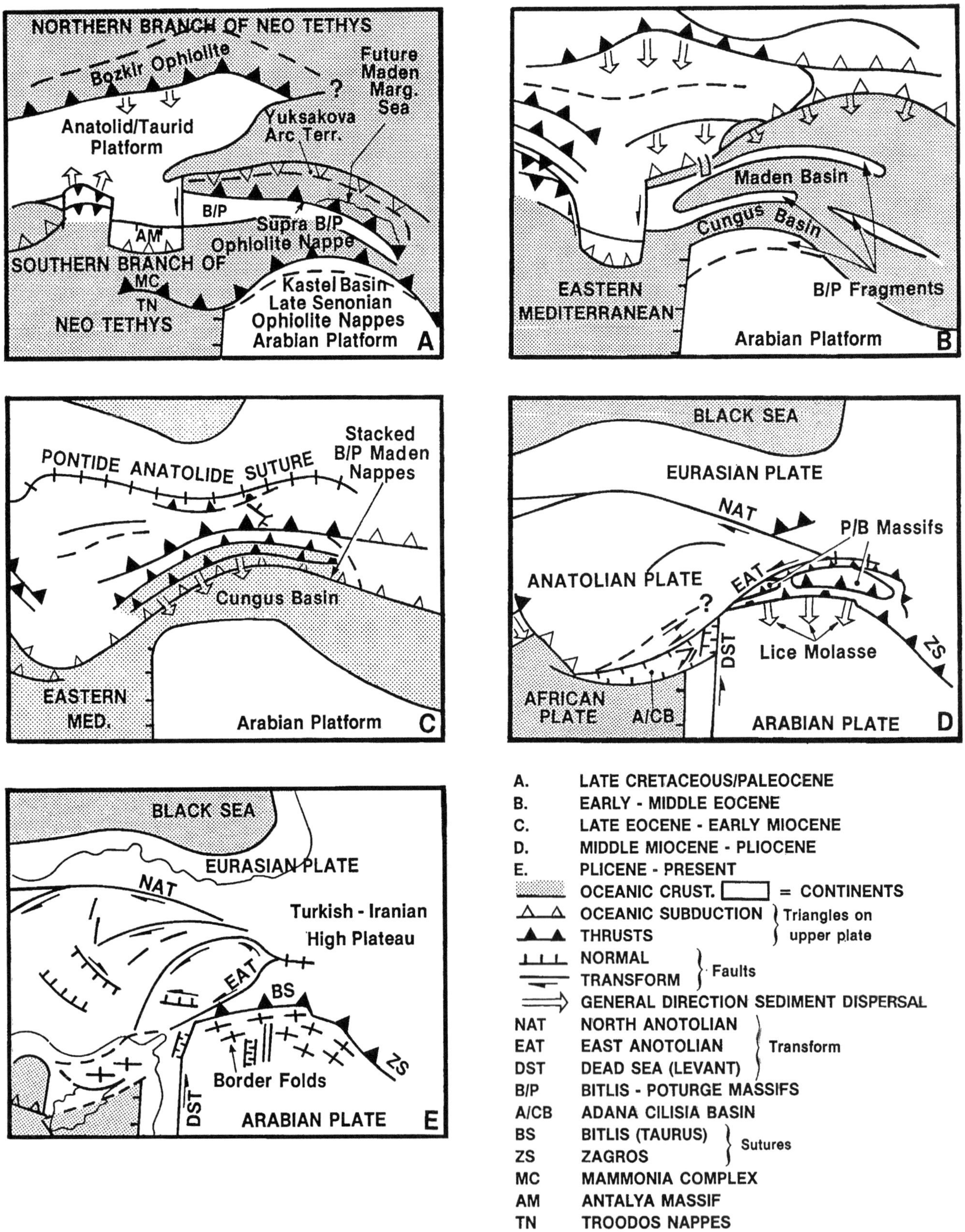

Figure 19—Cenozoic plate tectonic reconstructions: north Arabian margin and Turkey.

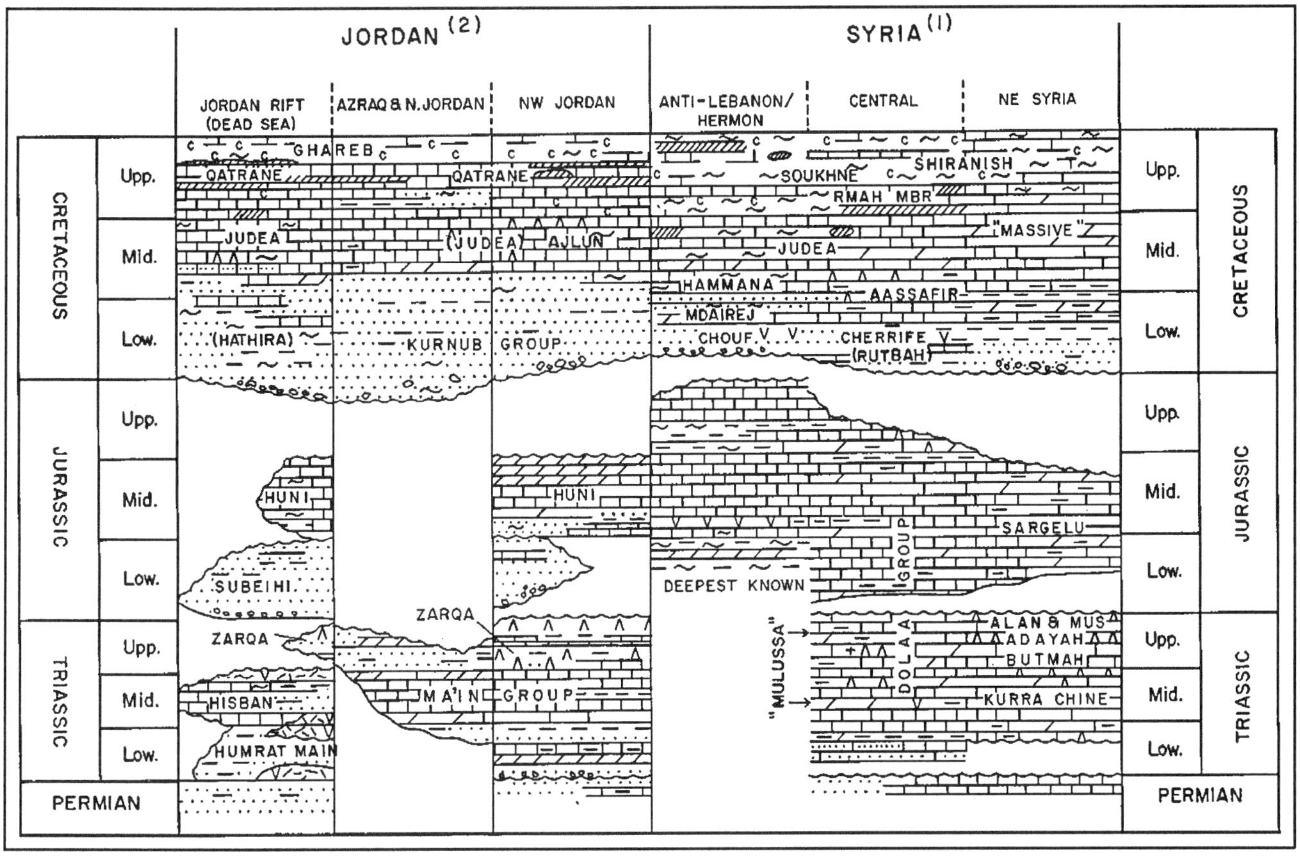

Figure 20—Mesozoic formations correlation: northeast Syria–Jordan. For legend see Figure 5.

developed in front of the advancing allochthons. Obduction onto the Arabian platform margin may have been the result of collision with the Bitlis-Poturge fragment and the ophiolites have a common latest Maastrichtian conglomeratic sedimentary cover (Sengor and Yilmaz, 1981). The Maden marginal complexes probably began about that time, that is, shortly after south dipping subduction had been initiated north of the combined Bitlis-Poturge-Arabian continent (Figures 19A and 19B).

The Maden and Cungus oceanic basins reached their maximum development early in the Paleocene, while the Yuksekova arc had ceased its activity. Eocene Midyat carbonates (Figures 14B, 21) were being deposited in the southern half of the Cungus basin, indicating a return to quiet shelf conditions along the northern margin of the Arabian platform (Sengor and Yilmaz, 1981).

By the late Eocene, the Maden basin had closed and the event interleaved Yuksekova, Maden, and Bitlis-Poturge lithologies and the Eocene flysch fill of the Inner Tauride ocean. Early in the Miocene, the Cungus basin began receiving large amounts of flysch sediments and olistostromes (Figure 19C). A sinuous north-dipping subduction zone beneath southern Turkey consumed the young oceanic(?) or quasi-oceanic lithosphere of the Cungus basin. During the early to middle Miocene (Serravalian) this basin closed, Arabian-Eurasian collision began along the Taurus (Bitlis) suture, and Lice molasse was deposited in the foredeep that arose on the Arabian platform to the south (Figure 19D). This was soon followed by terminal collision along the Zagros suture. The resultant north-south shortening, thickening, and elevation of the east Anatolian plateaus was eventually complimented and compensated by wedging out of a considerable portion of Turkey westwards along the east and north Anatolian transforms, thus giving rise to the Anatolid plate (Dewey et al., 1986) (Figures 19D-E).

The east Anatolian transform and the current, new northwest Arabian plate margin delineated by the Levant (Dead Sea) transform, meet at an FFF triple junction to form the Miocene Adana-Cilisia intra-continental basin. This basin resulted from incompatibility problems arising in such a continental triple junction and is located at the northern end of the Sinai-Levant continental fragment between the old northwest Arabian margin and the new one to its east. Continental convergence of Arabia with Eurasia in the Pliocene gave rise to the border fold region on the Arabian platform as a foreland fold and thrust belt area. This continental convergence also fostered the north-south struc-

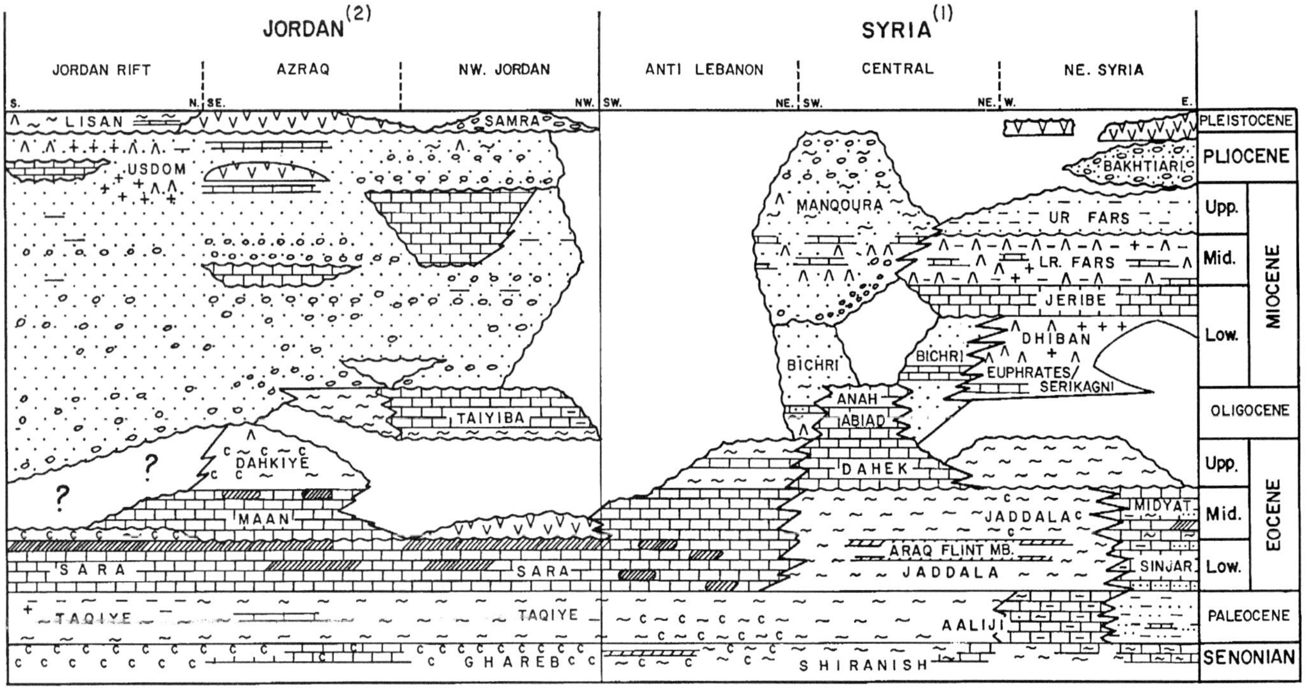

Figure 21—Cenozoic formations correlation: northeast Syria–Jordan. For legend see Figure 5.

tural features in southeast Turkey such as the large volcanic region of Karacalidag, as impactogens (Sengor and Yilmaz, 1981).

On the basis of additional mapping within the Bitlis suture zone, combined with age control of suture zone lithologies and use of data by Dercourt et al. (1986), Hempton (1987) suggested that initial suturing between Arabia and Anatolia occurred in the middle to late Eocene and terminal suturing occurred by late in the Miocene. The initial suturing was produced by convergence between cold, stretched and thinned Arabian leading edge continental crust and a hotter composite Anatolian-Eurasian plate composed of arc, continental, and ophiolite terranes. The thrust-stacking of suture lithologies commenced with this initial suturing. The north-south so-called Dead Sea (Levant) fracture was initiated shortly thereafter in the late Oligocene–early Miocene, as initial extension in the Red Sea/Gulf of Aden began and left-lateral movement along this new Levant fracture developed. This opened up the pull-apart basin of the Dead Sea, supplanted the Mediterranean margin as the current, new northwest Arabian margin, and commenced deformation of the Palmyrides in Central Syria. Sinistral movement ceased by mid-Miocene allowing the Dead Sea and other "pull-aparts" on trend (?the Ghab) to fill with sediments. Horizontal motion was resumed early in the Pliocene leading to further opening of the pull apart basins and further deformation and uplift of the Palmyrides during underthrusting of their southern margin.

In the hinterlands to the northwest and north margin areas, a series of intra-plate, fault-controlled basins occur that developed as a consequence of different phases in the structural evolution of the two margins. These include the Palmyride basin and the Euphrates graben with its differently oriented segments. The Palmyride basin (Figure 1), trending northeast-southwest over a 350 km length and 100 km width, originated in the Early Triassic together with the evolution of the Levantine northwest margin during the Late Permian–middle Triassic rifting events discussed above. It evolved as an aulacogen or failed rift (Ponikarov et al., 1967; Beydoun, 1981) connected to that margin (Figure 1), and is now inverted as a deformed Mesozoic/Cenozoic belt sandwiched between two relatively stable blocks: the Rutbah uplift to the south and southeast (a segment of the regional Ha'il- Ga'ara [or Rutbah]-Khleisia–Mardin high), and the Aleppo high or platform to its north. Both of these stable blocks have thick Paleozoic and thin Mesozoic–Cenozoic successions and probably represent broad inverted former Paleozoic intra-cratonic subsident platforms characteristic of much of the Arabian plate after cratonization. By contrast, the Palmyrides accumulated at least 5,000 m of post-Paleozoic sediments, thickening abruptly into the basin from the arched Paleozoic highs (McBride et al., 1990). These authors considered that the southern platform (Rutbah uplift) was active in the early formation of the Palmyride trough, with downflexuring of the lower Paleozoic platform sequence beneath the southern front of the Palmyride belt along a basement hinge zone as early as the Carboniferous. Simultaneously, the Aleppo high acted passively. No major basin-bounding rift faults can be identified from the seismic data.

The Palmyride basin was partly inverted in Late Cretaceous time, but the basin's principal inversion and deformation was in the Cenozoic. This was by the combined effect of left lateral Miocene movements along the new northwest Levant margin (Dead Sea) transform boundary and the Bitlis-Poturge/Arabian plate collision from the middle Miocene on (Beydoun, 1988).

Hempton (1987) saw the south-directed thrusting of the southern Palmyride fold belt as a consequence of terminal suturing of the Arabian/Eurasian plates, with northward Arabian plate movement resuming in the Pliocene as a consequence of Red Sea/Gulf of Aden spreading in the south, and of Anatolian plate wedge escape westwards in the north. The southern Palmyride zone thus became a focus of a new incipient north-dipping subduction zone (Rothstein and Kafka, 1982). Lovelock (1984) noted that the Palmyride zone is too wide and extensive to be simply a belt of en echelon folding associated with the left lateral motion on the Levant (Dead Sea) transform as suggested by Beydoun (1981). Lovelock inferred the deformation to be in response to ridge push forces originating in the Red Sea that were transmitted to the eastern or Arabian block. As this block began to compress as a result of collision with Eurasia, the most readily deformable portion of the northern Arabian plate would have been the Palmyride trough with its several thousand meters of undercompressed post-Paleozoic sediments, including Triassic evaporites. This compression engineered the Palmyrides underthrusting and box folding from the middle Miocene onwards.

The nearly west-east trending Jhar fault, with its strike-slip and thrust components, divides the Palmyrides into two distinct provinces. The southwest part owes its en-echelon conjugate folds mainly to compression by left lateral movement on the Levant transform. In contrast, the northeast segment, with its greatly thickened Mesozoic section, may have formed at the site of a zone of intra-plate weakness that took up the excess shortening between two rigid blocks of the Arabian plate (McBride et al., 1990).

During the Coniacian-Santonian, and 25 Ma before the ophiolite obduction in the late Campanian-early Maestrichtian along the edge of the north Arabian margin, evidence of considerable rifting in intra-plate depressions of the northern Arabian plate is well documented. This includes the Palmyride and Sinjar troughs and along the Euphrates-Anah-Abu Jir graben segments of northern Syria and Iraq. There is a strong suggestion from the fault pattern that extension was primarily north-south, leading to sinistral oblique slip along the northwest-southeast-trending Euphrates and Abu Jir graben segments, although the amount of opening was relatively small (Lovelock, 1984). Lovelock attributed these events to the fact that the north Arabian passive margin was stretched as it approached a north-dipping subduction zone associated with the Bitlis-Poturge continental fragment (distinct from the south-dipping zone, to the north of these fragments, postulated by Sengor and Yilmaz, 1981). The Kastel foredeep that developed north of the northern Arabian margin became an active trench, into which ophiolite sheets and flysch were deposited by gravity as described by Rigo de Righi and Cortesini (1964). The rifting suddenly terminated as the direct result of the collision of the continental part of the Arabian plate with the trench, thus blocking subduction and returning the north margin to compression (Lovelock, 1984).

The Infracambrian and Paleozoic sedimentary succession of the north and northwest margins and their hinterlands are very similar in facies and in areal distribution to those already described for the Arabian sector of the northern Gondwanan margin and correlated in Figures 5 and 6. The formations encountered both in outcrop and in the subsurface in southeast Turkey, Syria, and Jordan exhibit a predominance of clastic facies consistent with the regime prevailing throughout the Arabian and adjoining plates. Carbonate depositional episodes are restricted to the Middle Cambrian and Late Permian; Devonian sediments are rare in this region. Along the Levant portion of Arabia, the Paleozoic and older successions are known only from the Amanus Range in the north and the Sinai/Negev region in the south, where they have been brought to the surface by later tectonics or are at relatively shallow enough depths to have been penetrated partially by the drill. In the Amanus region, the sequence forms part of that shown for southeast Turkey in Figure 5, while in the Sinai/Negev area the succession includes some Infracambrian and Cambrian-Ordovician clastics and Carboniferous-Permian clastics and carbonates (Beydoun, 1988). In the intervening Levant sector, the very thick Mesozoic-Cenozoic sedimentary cover has ensured that the underlying sequence has only been rarely and partially reached by the drill.

As it is over most of Arabia, the Mesozoic sequence of the north and northwest margins is predominantly carbonate but is much divided into packages by numerous regional unconformities, or more local breaks, especially for the north margin. These breaks reflect the effects of regional tectonics related to plate margin and intra-plate basin genesis and evolution. The succession for the southeast Turkey part of the north margin is shown in Figure 12 and is correlated with that of Iraq and southwest Iran on the northeast margin. Large areas of southeast Turkey have been stripped of much of their earlier Mesozoic sediments, with middle Cretaceous carbonates often lying on deeply eroded Paleozoic or even basement rocks as a result of the Late Jurassic-Early Cretaceous block movements discussed earlier. The lower Mesozoic deposits appear to be preserved only in the basins surrounding the major geanticlinal Buzova-Mardin uplift (Figure 1). This uplift developed both as a result of the early rifting and the subsequent Late Jurassic movements. It separated the Arabian shelf from the continental edge to the north during the Triassic through the Early Cretaceous, although it probably had a lower Mesozoic cover that was subsequently removed. To the north, east and south, the Triassic and Jurassic carbonates with shales and/or evaporites are preserved and represent the first main Mesozoic cycle. The second cycle commenced in the Albian with a transgression that overlapped and overstepped the old highs and left behind a series of reefoid carbonates (Mardin group). Post-Turonian movements led to differentiation of the platform areas and to differential subsidence of former uplifted portions, and culminated in the north with the formation of the narrow (subsidiary) Kastel foredeep, by the Campanian, which was then filled with flysch and the already-described gravity nappes and olistostromes (Rigo and Righi and Cortesini, 1964). The flysch, synchronous allochthonous complexes, and coeval autochthonous sediments are tabulated in Figure 12, and often cover older potential petroleum targets buried beneath them along the Arabian shelf edge of the north margin. Locally these gravity slides continued into the Paleocene.

The Mesozoic succession of the current, new northwest Arabian margin in Syria and Jordan is shown in Figure 20, and includes facies across the hinterland of the north margin in Syria and through the Palmyrides in central Syria to the current plate margin of the Anti-Lebanon and Jordan-Dead Sea rift. The successions are predominantly carbonate with subordinate localized evaporites such as the Triassic evaporite of northeast Syria, the Palmyrides and the Jordan rift edge, and are related to the initial rifting phase of Levant-Turkey separation. Following the Late Jurassic block uplifts and erosion, the Lower Cretaceous documents a widespread phase of continental clastic sedimentation, in places with associated extrusives, that is particularly well seen in Jordan. Marine conditions did not return until Aptian-Albian time, with shallow-water carbonate deposition in the middle Cretaceous and Upper Cretaceous pelagic carbonates. On this part of the Tethyan (Mesogean) southern margin, these late Cretaceous conditions are reflected by high organic productivity with extensive phosphorite-rich deposits all along the north and northwest margins and hinterlands of Arabia and North Africa.

Similar Mesozoic facies are recorded from the Levant-Sinai portion of the former Arabian plate, with the earlier Mesozoic (Triassic) being relatively unknown because of deep burial and little or no penetration (Lebanon) by the drill. In maritime Syria, the Triassic is predominantly in a dolomitic facies with no evaporites (Mouty, 1976) but the base is unexposed. Much of the Upper Jurassic section has been eroded in the northern Levant.

The Cenozoic sequences along the north and northwest margins are shown in Figures 14 and 21. That for southeast Turkey shows a complete sequence on the platform, predominantly carbonates (often reefal) with interfingering shale and some evaporite levels. In the Turkish part of the margin farthest north, the Eocene-Miocene phase was characterized by flysch/ molasse clastics with Miocene-Pliocene allochthonous units of the Ergani Complex, including ophiolites, thrust onto the shelf from the Anatolian collisional fragments (Bitlis-Poturge-Yuksekova, etc.) These include Paleocene-Eocene-Miocene rocks with Bitlis metamorphics, Guleman and Yuksekova complexes, Hakkari and Maden complexes and the young Cungus formation over which the Selmo Pliocene molasse formation lies flat. The Lice and Selmo formations are the Miocene-Pliocene units of the flysch-molasse that covered the north margin, and were overlain in the Pliocene-Pleistocene by the basaltic flows of the Karacalidag and other volcanos.

Along the current, new northwest margin in Syria and Jordan, including the Palmyride and other intraplate basins of the region (Euphrates, Azraq, Dead Sea, etc.), the Paleocene was characterized by pelagic open marine deposits with organic rich levels. The Eocene was principally a time for shelf carbonates with marls characterizing the Palmyrides.

Late Eocene uplift of much of the region, with initial Arabian leading edge collision of the Bitlis and Zagros sutures and crustal restacking, left only a shallow seaway between the Mediterranean Sea and the Indian Ocean across northern Syria and Iraq that continually receded. Inland, the Jordan-Levant and Azraq rift areas acquired continental clastic fills starting with the Oligocene, interrupted by lacustrine and limited periodic marine incursions culminating in thick evaporites in the Dead Sea pull apart from latest Miocene-Pliocene time to the present.

Along the Sinai-Levant portion of the former northwest margin, the Late Eocene-Holocene deposits were restricted to the coastal region following late Eocene uplift of much of the area. These deposits consist of thick organic-rich marls and evaporites including well-developed Messinian evaporites in the southern Levant, and much more areally restricted Miocene-Pliocene carbonates, clastics, and volcanics in the northern Levant. The exception is the Kabir River Valley near Lattakiya where a more complete section is found, including Messinian evaporites linked to the Mediterranean salinity crisis.

Chapter 7

The Neogene Split From Africa: The Arabian South and Southwest Margins

The Arabian south and southwest passive margins evolved during the Neogene as a result of Eocene-Oligocene extension, rifting and breakup of Afro-Arabia. Separation of Arabia away from Africa followed as a consequence of sea-floor spreading in the Gulf of Aden. This spreading propagated to the Red Sea and rotated Arabia in a counter-clockwise direction, simultaneously moving it northward along the Levant Fracture (Dead Sea transform), which became the new northwest Arabian plate margin.

The Gulf of Aden opened initially, it appears, because of the westward propagation of the spreading Indian Ocean Carlsberg Ridge. This propagation was accompanied by an increasing component of compression developing along the Owen Fracture Zone (OFZ) transform in response to a sharp change in spreading direction during the earlier Eocene (Moseley and Abbots, 1979). The Sheba Ridge of the Gulf of Aden only developed when sea-floor spreading became established in the Miocene, and it constitutes a dextrally offset extension of the Carlsberg Ridge on the west side of the OFZ. By contrast, as yet the Red Sea has no central ridge but has only an axial trough principally in the southern and central portion, which is a current site of high heat flow (Figures 1 and 24).

Researchers have proposed a number of models for the evolution of the Gulf of Aden and the Red Sea. To date, these researchers dispute such points as whether, once sea-floor spreading commenced, it was episodic or continuous; whether the crust under the Red Sea is oceanic across much of the basin as is the case in the Gulf of Aden, or only in the discontinuous axial fissure; whether the basin is asymmetric, with oceanic crust on one side and only stretched continental crust on the other; whether the floor of the intervening Afar depression (Ethiopea-Djibouti) is principally oceanic or attenuated continental crust; whether rifting was initiated after thermal doming and uplift, with resultant volcanicity and subsequent axial collapse, or developed as a passive rift without doming; and whether the Red Sea (and for some, also the Gulf of Aden) formed along a shear belt, tearing open the Gulf of Aden. I reviewed most of these disputed views recently (Beydoun, 1988, 1989a), and I summarize the essential observations below.

The Gulf of Aden and the Red Sea have constituted zones of crustal weakness and instability in the Nubo-Arabian and Somali shields since Pan-African cratonization. Marine ingressions occurred intermittently(?) onto these zones of weakness during the Mesozoic and early Paleogene (and probably in the Paleozoic, although no record remains) as narrow sea-arm invasions when the bordering land was of low relief and close to sea level.

Marine Jurassic formations are present in outcrop on either side of the southernmost Red Sea and on both sides of the Gulf of Aden and have been encountered in offshore wells drilled on the Arabian side of the latter. In terminal Jurassic/Early Cretaceous time, roughly north-south-trending block uplifts stripped Jurassic sediments from large positive horsts, but they were preserved in troughs (Beydoun, 1982, 1988). There were also times when regional tectonic stresses rejuvenated older northwest-southeast Najd fractures and propagated a rift system across southwest Arabia-Somalia. These rifts were filled with Upper Jurassic-Lower Cretaceous sediments in the Marib-Shabwa-Balhaf area of Yemen. A probable triple junction? may be present near the Shabwa area of the northwest-southeast trending Marib-Shabwa-Balhaf rift. This junction has an apparent offshoot to the east along the Hadhramaut valley depression that links with the Jezaa trough lying south of the Hadhramaut arch (Figure 1). The Jezaa trough may have been initiated at the end of the Jurassic and subsided greatly during the Cretaceous, having received a considerable thickness of Cretaceous sediments—predominantly marine carbonates and argillaceous deposits with source rock potential (Haitham and Nani, 1990; Paul, 1990). A third structural depression, the Sir basin, may be part of the triple junction linked to the Marib-Shabwa-Balhaf failed rift or, more likely, may be an offshoot fault-bounded depression. The geophysical data published are few and do not make a more confident conclusion possible (see summary data in Paul, 1990).

Marine Upper Cretaceous-lower Paleogene sediments occur sporadically in the northern half of the Red Sea and are preserved in down-faulted blocks in the shield on the Arabian side. Such sediments are also encountered in the subsurface of onshore wells drilled in the Sudan (see Beydoun, 1989a for summary) (Figures 17, 18, 22, 23).

Uplift, extension, and rifting on the trend of the present margins did not commence in the region until the

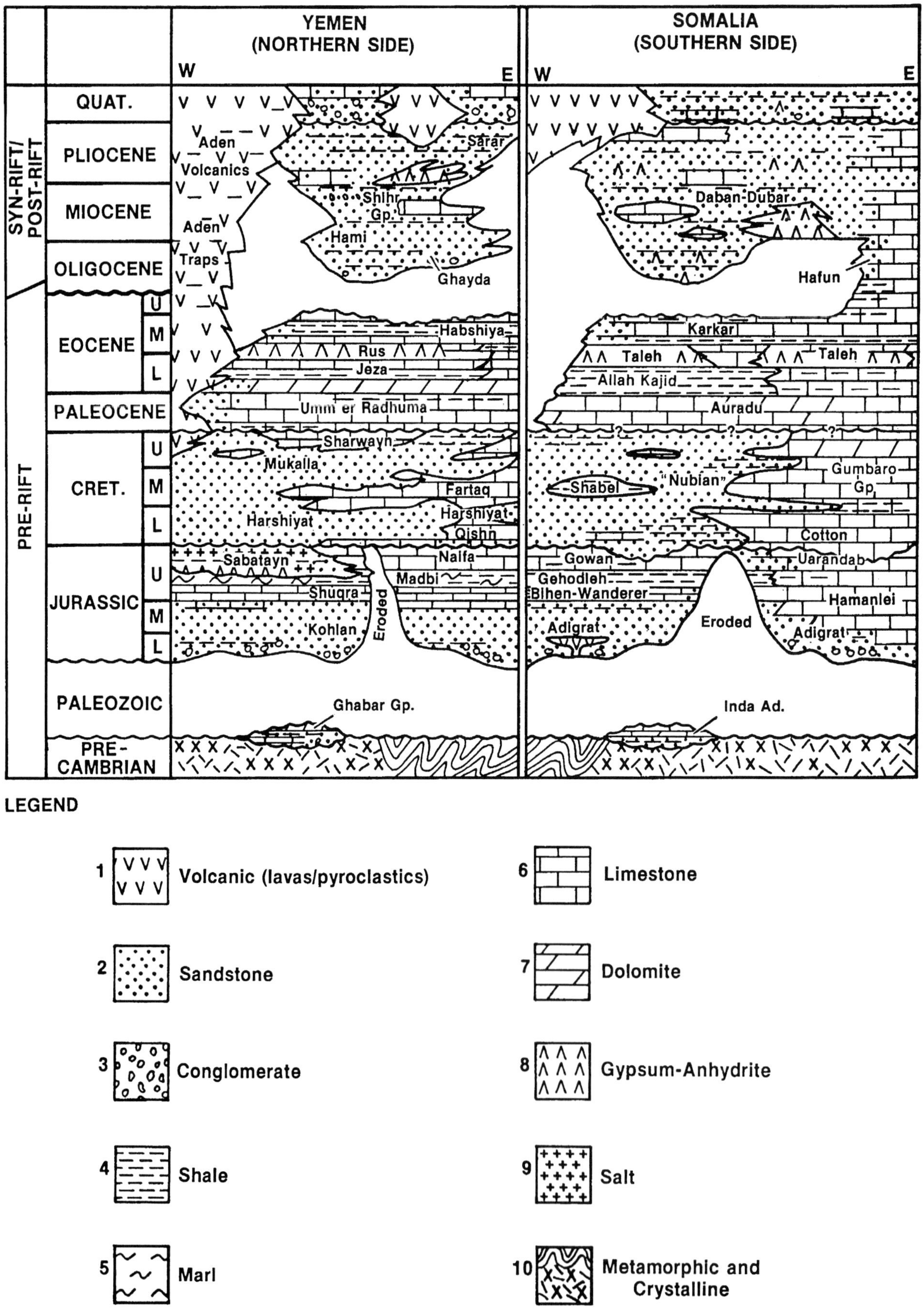

Figure 22—Phanerozoic formations correlation: Gulf of Aden basin.

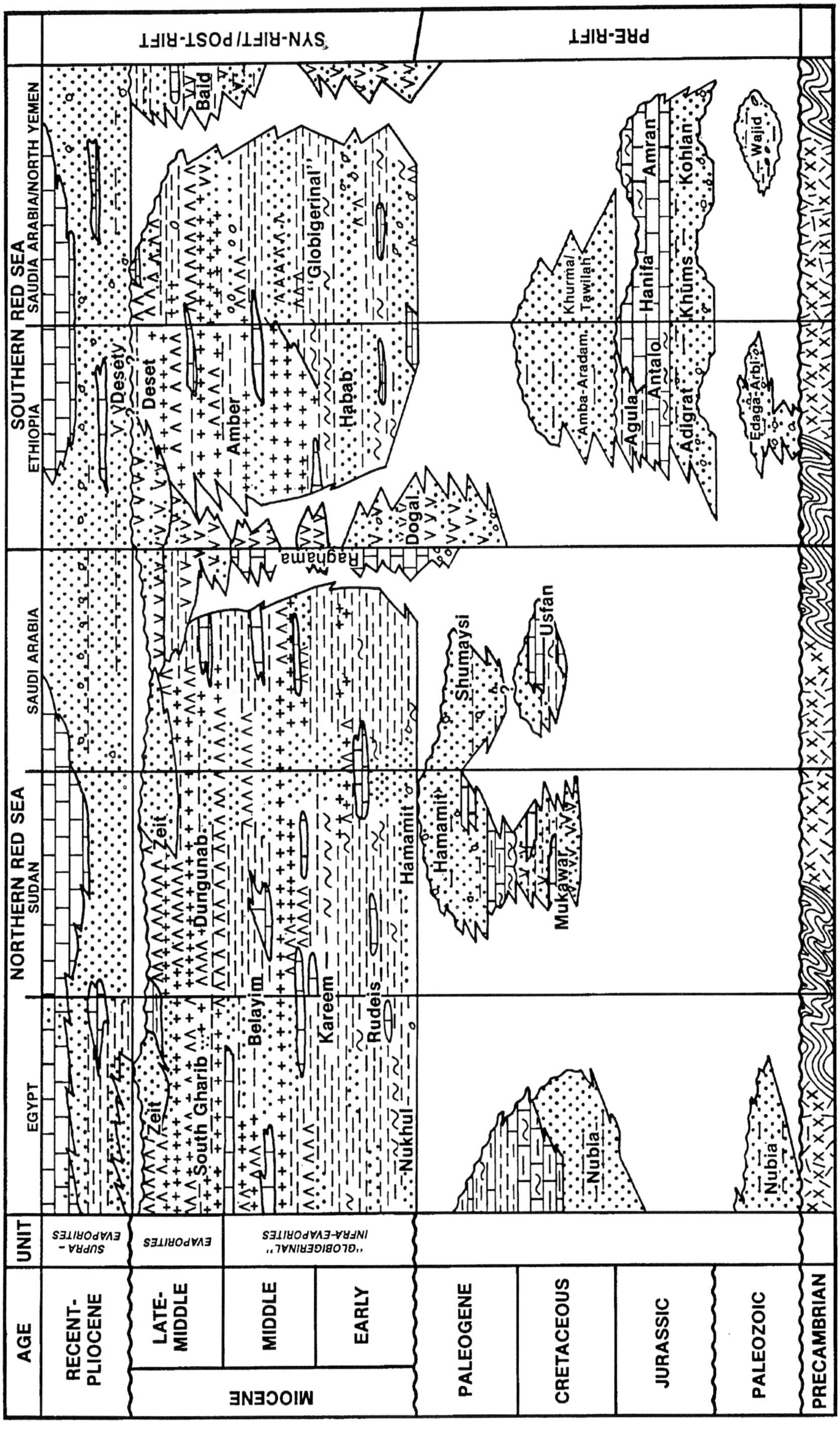

Figure 23— Phanerozoic formations correlation: Red Sea basin and hinterland. For legend see Figure 22.

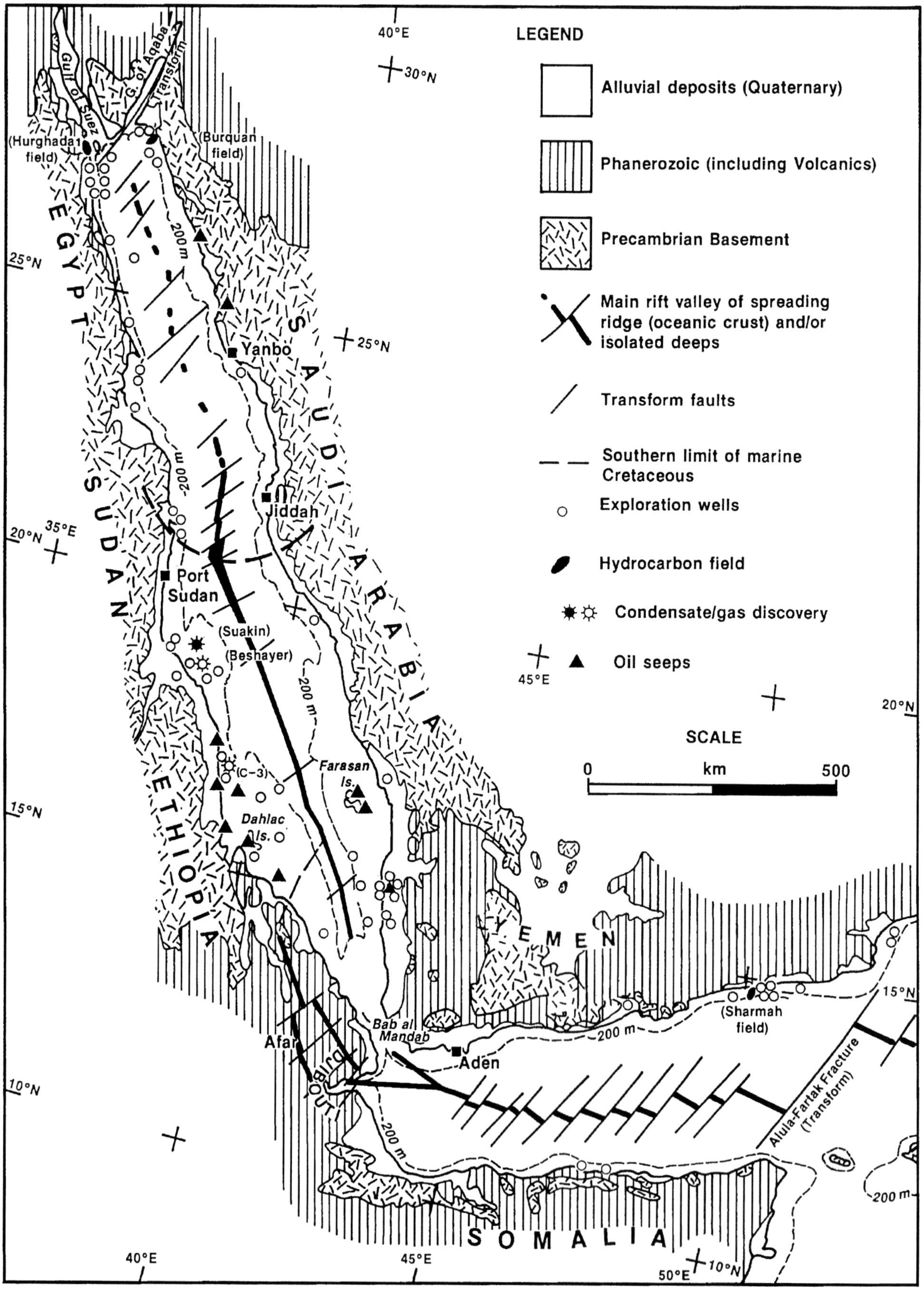

Figure 24—Red Sea–Gulf of Aden regional map.

Eocene-Oligocene and became fully developed during the Neogene. This marked the separation of Arabia from Africa for the first time since their formation in the late Proterozoic Pan-African cratonization. Events leading to this separation probably were episodic, occurring in two stages separated by a tectonically quiescent stage as proposed by Laughton (1966), Girdler and Styles (1974), Girdler et al. (1980), Girdler (1985), Girdler and Southern (1987) and others. Hempton (1987), however, preferred an earlier initiation of active separation than is otherwise agreed, although he considered that active spreading did not begin in either basin until 5 Ma (Hempton, 1987; Berhe, 1986). Cochran (1981, 1983), Bonatti (1985), La Breque and Zittelini (1985), and others who favored continuous separation, prefer 10 Ma as the initiation of active sea-floor spreading. Bonatti (1985) also proposed that initial emplacement of oceanic crust in the axial fissure is in regularly spaced hot points that then serve as nuclei for axial propagation into segments of oceanic crust. His suggestion was made in order to explain the discontinuous segments of axial trough in the northern Red Sea.

Confusion arises out of opposing claims for oceanic versus stretched continental crust underlying much of the Red Sea and even under the Arabian coastal plain (Gettings et al., 1986; Girdler and Underwood, 1985). This interpretation also supports an earlier phase of sea-floor spreading than is generally accepted and is postulated to begin in late Oligocene by Styles and Hall (1980) for the whole of the Red Sea-Gulf of Aden. The dispute may be resolved by recently acquired geophysical data including deep, wide-angle reflection and refraction seismic in offshore Egypt, Sudan, Saudi Arabia, and Yemen (Makris and Rihm, in press; Richter et al., in press). Modelling that uses these deep seismic data with gravity and magnetics indicates that the Red Sea basin is asymmetrical with oceanic crust underlying much of the African side and only stretched and thinned continental crust underlying the Arabian side. It is inferred that no true sea-floor spreading, other than that recently occurring on the axial fissure, had taken place. The oceanic crust under the African side is not due to an earlier spreading phase, but is interpreted to be the result of initiation and coalescence of a series of en echelon pull-apart basins that developed in response to shear on the Red Sea prolongation of the Levant (Aqaba-Dead Sea) transform. These pull-apart basins were initiated some 14 Ma because acceleration of Arabian plate separation movement in the Miocene could no longer be accommodated by the opening of the Gulf of Suez (see also Beydoun, 1989a). These initially north-south morphotectonic depressions (Coleman, 1984) had become a single large lake by the late Oligocene-early Miocene and received lacustrine sediments. However, although this area had become a marine flooded basin early in the Miocene with incursion from the Mediterranean, still it was thought to be mainly cut off from the Gulf of Aden. Recent biostratigraphic work, however, (Hughes, et al., in press) indicates that marine connection with the Gulf of Aden had already taken place in the ?earliest late Oligocene in the southernmost Red Sea and in the central and northern Red Sea during the early Miocene. This lends support to the concept of northward propagation of opening proposed by Courtillot et al. (1980). The marine connection between the Gulf of Suez and the Mediterranean Sea does not seem to have occurred until the early—perhaps earliest—Miocene. Gillman (1968) suggested the possibility of intermittent Oligocene connections between the Red Sea and the Indian Ocean across Afar while he still subscribed to the general view that the main link with the Indian Ocean was of Pliocene age.

The sedimentary successions of the Gulf of Aden and Red Sea basins and of the adjacent Arabian and African margins can be divided into three main groups: (a) the prerift sequence; (b) the synrift sequence; and (c) the postrift and drift sequence.

The Gulf of Aden sequences correlated from the Arabian and African sides are shown in Figure 22. More detailed correlations with adjacent and other parts of the Arabian plate, particularly with Oman and the Red Sea part of former North Yemen, are shown in Figures 6, 17 and 18.

The older prerift succession is only sporadically preserved in outcrop and imperfectly known from the subsurface. A series of folded and slightly metamorphosed sedimentary rocks consisting of two clastic and two carbonate formations comprising the Ghabar group is preserved in down-faulted basement blocks near the coast in Yemen. The series is equated with the Huqf group of Oman and the Inda Ad series of Somalia (Beydoun 1964, 1982, 1988, 1989b). Inland along the south margin, Paleozoic clastic sediments of possible Cambrian-Ordovician and Silurian-Devonian ages onlap the Hadhramaut arch from the north (subsurface sediments of the Rub al Khali basin) and onlap the Yemen uplands from the east. The Wajid formation, of disputed age, crops out in the Yemen and Asir uplands of southwest Arabia. Its base is glaciogenic and has been separated into a distinct "Akbara" unit (Kruck and Thiele, 1983; El-Nakhal, 1984) that can be correlated with the Haushi group Al Khlata formation of Oman, of Late Carboniferous-Early Permian age. This places most of the Wajid as being Permian in age. I discussed these age problems of the Wajid in more detail elsewhere (Beydoun, 1988, 1989b). In the northwest-southeast Marib-Jauf graben (Figure 1), Wajid and "Akbara" and older clastics of Ordovician age are present in the subsurface (Maycock, 1987, oral communication).

The Mesozoic sequence of the Gulf of Aden commences with the Kohlan formation, a transgressive clastic sequence of Liassic-Middle Jurassic age that filled hollows in and then covered the peneplaned basement (or Paleozoic deposits where preserved). This was followed by shallow and then open marine carbonates of Late Jurassic Callovian-Tithonian age that were partly contemporaneous with evaporites, including salt, equivalent to the Hith evaporitic formation of Late Jurassic age that was deposited extensively over the northeast Arabian plate margin (Beydoun, 1964, 1988). The salt was deposited in the inland northwest-southeast rift basins of Yemen, which developed in Kimmeridgian time (Maycock, 1987, oral communication) and which include the major Marib-Jauf-Shabwa-Balhaf rift, cutting the South Arabian coast (Figure 1). This rifting episode, on rejuvenated Infracambrian Najd fractures, was followed in the latest Jurassic–Early Cretaceous by block uplift movements generally trending north-south. Erosion stripped the Jurassic sequence from the elevated blocks. The Cretaceous transgression advanced from the opening Indian Ocean in the east and was dominantly carbonate in the eastern part of the south margin and principally continental clastic deposits in the west. The end Cretaceous events that culminated in ophiolite obduction along the north, northeast, and southeast margins had their counterpart in the Gulf of Aden-Red Sea region. The counter events included the start of thermal doming of the Ethiopian-Yemeni region and the commencement of Trap basalt outpourings, which culminated in late Paleogene-Neogene time, and a short regression of the sea to the east. The Paleogene began with the powerful Umm er Radhuma transgression, which deposited carbonates and some basal shales over almost the whole of eastern and southern Arabia and extended well into Somalia. This transgression and deposition event was followed by shoaling with widespread evaporite deposition (Rus formation) followed by a return to shallow-marine carbonate/shale deposition in the middle Eocene. Regional uplift of much of Arabia occurred in the latest Eocene and was accompanied by erosion.

Gulf of Aden rifting and subsidence commenced in the Oligocene followed by breakup and separation of Arabia from Africa developing in the Miocene and giving rise to a thick, varied, synrift succession. This sequence is predominantly clastic but includes some evaporites and carbonates that reach 2000 m in thickness in the offshore area. The sequence is collectively known as the Shihr group of Oligo-Miocene age (Beydoun 1964, 1988, 1989a; Beydoun and Greenwood, 1968), and also includes brackish and episodically hypersaline marginal marine facies of early to middle Oligocene age (Hughes et al., in press). The Shihr group has recently been subdivided into recognizable formations as a result of offshore drilling data (Haitham and Nani, 1990). Shihr sediments and equivalent deposits on the African side ingressed into the subsiding and opening Gulf of Aden and merged upward with similar postrift deposits of Pliocene age that represent the drift stage. Onshore the Shihr sediments are found only in outcrop in low-lying areas bordering the Gulf of Aden. Farther west, in the Aden and Yemen plateau areas, coeval deposits consist principally of the volcanics of the Aden Trap and Aden Volcanic series (Beydoun and Greenwood, 1968).

In the Red Sea region that is now the southwest Arabian margin, prerift Jurassic successions are present only in the southernmost part exposed on either side but not reached by the drill offshore. The deposits are similar to those of the Gulf of Aden region and the highland of Yemen but without the evaporites, and contain, on the Ethiopian side, organic-rich shales (Figure 23). The underlying clastic Paleozoic sequence is regarded as being of Carboniferous-Permian age. Post-Jurassic prerift deposits are restricted to outcrops on either side and are predominantly coarse clastics with volcanics. Carbonates and shales of Late Cretaceous/early Paleogene age have been recorded in the northern half of the Red Sea in down-faulted blocks near Jeddah (Karpoff, 1957) and in the subsurface of onshore wells on the African side in Sudan (Carella and Scarpa, 1962). The remaining prerift sequence is clastic but in the Red Sea area of Egypt the Late Cretaceous-Paleogene is dominated more by carbonates with marls and shales. Offshore, however, none of the relatively few wells drilled in the Red Sea (Figure 24) has encountered this sequence although one borehole reached the basement.

The synrift succession of the Red Sea basin (southwest margin) is very similar to that of the Gulf of Suez and consists of a pre- or infra-evaporite sequence of essentially late Oligocene-middle Miocene age, an evaporite sequence, including thick halite, of mid-late Miocene age, and a post- or supra-evaporite sequence of Pliocene-Holocene age that consists of marine clastics and constitutes the postrift/drift phase succession. Toward the rift basin edges, coarse clastics of Oligocene-early Miocene age fan out and grade into finer sediments in the center of the basin and are interbedded with volcanics associated with the rift fracturing. The infra-evaporite series contains several shale and marl levels with intercalated sands and occasional anhydrite beds and carbonate "stringers," which are collectively called the "Globigerinal Marl group." The latter correspond to the Gharandel group of the Gulf of Suez. This succession, essentially Miocene in age, succeeds the initial rifting sequence disconformably and represents restricted-circulation anoxic deposits of lacustrine to marine origin deposited as the sea invaded deeply into the subsiding basin. The sea's invasion was principally from the Mediterranean, but there apparently also were (?sporadic) connections with the Indian Ocean via the Gulf of Aden and Afar (see earlier). Many of these shales and marls constitute potentially rich source sediments and often are interspersed with generally good reservoir sediments. The evaporite group consists predominantly of thick salt with subordinate anhydrite, marginal sands, and some carbonates. The supra-evaporite group comprises terminal Miocene clastics with subordinate anhydrite beds and Pliocene marine sands and shales with intercalated reef carbonates from the drift phase that finally linked the Indian Ocean and Mediterranean Sea. Organic-rich shale levels occur in these sediments. The syn- and postrift sequences can reach well over 3000 m in thickness. Good correlations with Gulf of Suez units can be followed over the Red Sea basin. Current work for the UNDP/World Bank Red Sea-Gulf of Aden Regional Hydrocarbon Study Project (UNDP/RAB/88/021) is refining the biozonation and depositional environments as well as the source rock geochemistry of the region.

Chapter 8

Development of Other Structural Basins Within the Arabian Plate

The development of several of the intra-plate basins located close to the various Arabian margins has already been considered. These include the Palmyrides basin and the various segments of the Euphrates graben system, the Levant (Dead Sea) Fracture "pull-aparts," all within the north and northwest margin areas, and the Marib-Jauf-Shabwa-Balhaf graben system and the Jezaa trough and Sir basin of the southwest and south margin areas.

The remaining major intra-plate basins include the rifted Jafr and Azraq/Sirhan basins and the Risha basin in the Jordan/northwest Arabia region, and the Rub al Khali, Widyan and Tabuk basins of interior Arabia (Figure 1). The last four originated as intra-cratonic sag basins mainly of Paleozoic origin, several of which were subsequently inverted and consolidated as structural highs.

The Jafr basin is a fault-bounded north-northeast–south-southwest-trending structural depression with over 4000 m of sediments in the axial portion below the Middle Cambrian Burj limestone formation (NRA, 1986). The sequence was assumed to represent Lower Cambrian deposition over the basement but is here considered to be mainly equivalent to the Infracambrian/Lower Cambrian succession that accumulated in extension–related tectonic basins elsewhere in Arabia, mainly the Najd fracture and Oman salt basins, which permitted extra-thick syn-subsidence accumulations to occur and be preserved. A recent exploration well targeted to explore this sequence was abandoned at 4570 m as a dry hole, reportedly because of lack of source and seal sediments in a predominantly clastic sequence, but the presence of overlying organic-rich Silurian shales both here and elsewhere in Jordan (NRA, 1986) precludes dismissal of the Paleozoic prospects of this basin and the adjacent Sirhan-Tabuk basins.

The Risha basin accumulated a complete and thick sequence of Paleozoic sediments and was then inverted and angularly truncated by erosion to become part of the Ha'il-Ga'ara axis. The same can be said for other components of this wide axis, including the Rutbah and Khleisia highs and the Aleppo high of northwest Syria, all of which have thick Paleozoic sequences with Silurian shales. These highs were at times tilted or truncated and/or eroded, however, and will require better subsurface control to constrain potential source distribution.

The Tabuk and Widyan basins of northwest Saudi Arabia and northeast Saudi Arabia/Iraq on either side of the Ha'il-Ga'ara (Rutbah) arch are of probable Infracambrian origin, and have accumulated more than 5000 m of Paleozoic sediments. The Paleozoic sequence is complete in the Widyan basin but the Carboniferous and Permian successions are missing in the Tabuk basin (Al-Laboun, 1986). The influence on sedimentation of the intervening Ha'il-Ga'ara arch was minimal during subsequent early Paleozoic subsidence, but its rejuvenation in Late Devonian-Early Carboniferous time determined erosion/sedimentation rates all along the length of the arch to northern Syria/western Iraq. Rejuvenation again occurred in the Late Cretaceous and in the Paleogene and Pleistocene when regional tension permitted outpourings of basalt from fissures near the arch's crest in Saudi Arabia and other en echelon arches to the south and north (Greenwood, 1973). Potential organic-rich Silurian and Devonian source rocks are present in both basins as are potential reservoir-quality sandstones (Al-Laboun, 1986).

The Rub al Khali basin has generally been considered to be primarily a Tertiary epicratonic basin (Powers et al., 1966). However, the basin apparently originated in the early Paleozoic as an intra-cratonic sag basin and started to subside later in the Paleozoic (Sykes and Abu Risheh, 1989; Beydoun, 1989b). A moderately thick sequence of at least 1500-2000 m of Paleozoic sediments and 3000-8000 m of total sediments has accumulated (Figures 8 and 9). Silurian source sediments were deposited extensively over the basin, as were older and younger Paleozoic reservoir rocks.

Lastly but by no means exhaustively, on the list of intra-plate structural basins, is the Central Arabian graben and trough system located to the north of the east-west-trending Central Arabian arch (Figure 1). The arch acted as a hinge zone between a less mobile southern province and a subsiding one to the north. It was mildly positive early in the Paleozoic, but strong upwarping occurred in post-Early Devonian to pre-Late Permian time and subsided in the Mesozoic. The grabens and troughs are located where maximum tension along this hinge was directed. They apparently originated in mid- or Late Cretaceous time, with continuing subsidence in the Paleogene, in a tensile stress field generated by north-south stretching of the cover above an arcuate fault zone cutting the basement (Kadhi and Hancock, 1980). Movement along the system stopped in the Eocene, and a thick sequence of Cretaceous-Paleocene sediments is preserved in these depressions.

Chapter 9

The Hydrocarbon Potential of the Arabian Plate

WHY SO RICH IN OIL AND GAS?

The reasons for Middle East (Arabian plate) hydrocarbon abundance, particularly in the main producing basin(s) of the northeast margin (Enclosure 1), have been discussed by many workers specializing in the region. They include Henson (1951), Baker and Henson (1952), Lees (1951, 1953), Law (1957), Weeks (1958), Dunnington (1958, 1960, 1967, 1974), Kamen-Kay (1970), Halbouty et al. (1970), Kent and Warman (1972), Beydoun and Dunnington (1975), Grunau (1977), Riche and Prestat (1979), Murris (1980), Bois et al. (1982) and Beydoun (1986). I recently resummarized them and analyzed the differing views (Beydoun, 1988).

There is little doubt that one of the most important reasons is the fact that prevalent, large areas of high-quality source rocks have been preserved from different intervals of geologic time during the Phanerozoic, and especially during the Mesozoic, over much of the areally extensive northeast shelf, because of high organic productivity related to climate and oceanic circulation (Table 2). The development of juxtaposed source and reservoir facies belts was advantageous, as Dunnington (1974) indicated, because it facilitated expulsion and migration over long distances. Late Tertiary Zagros folding fractured inefficient Mesozoic cover and, aided by reversal of the gradient toward the rising fold front, also later facilitated vertical migration into better, younger, higher reservoirs beneath an excellent Miocene regional evaporitic seal.

Grunau (1977) considered that the main reasons for the petroleum richness of this region—more or less the northeast Arabian margin—lie in the extensive areas of mature source rocks, the huge volume of interconnected pore space within effective closures of huge traps, and the presence of major, effective, regional evaporitic seals that developed in Cambrian, Permian-Triassic, Late Jurassic and Miocene times (Figure 17). Multiple source–reservoir–seal development and a widespread sedimentary mega-sequence several thousand meters thick and preserved in the right sequential order, complete the picture.

According to Bois et al. (1982), the exceptional petroleum richness of this region is linked to its position on the passive margin of the latitudinal Tethyan realm; its voluminous reserves were generated by extensive organic-rich source rocks and an exceptional development of sediment volume related to several episodes of basin evolution linked with repeated ocean opening and closure.

However, Murris (1980) identifies the most outstanding factor of all. He rightly points out that the Middle East source rocks are not exceptionally rich or thick compared with those of other basins; neither are the Middle East's reservoirs, although excellent, in a class apart; nor are the seals, although effective, unusually thick or abundant. The horizontal scale of the basin (the northeast margin), however, is extraordinary, with the pre-erosional Mesozoic depositional platform being several thousand kilometers in length, around 2000 km wide and practically flat. This situation resulted in source, reservoir, and seal formation of very large areal extent and allowed very efficient horizontal migration that drained large hydrocarbon kitchens into extraordinarily large, often gentle structural closures. The structural gentleness of the Arabian region additionally allows only minimal loss of seal efficiency through fracturing and also provides large trap volumes for even modest vertical closure or oil column.

The extraordinary horizontal scale and the phenomenal width of the northeast margin platform are well illustrated in Figure 25. The figure was reproduced from Hughes Clarke (1990, personal communication [written]) with some small modifications/additions to show the extent of the northeast margin hydrocarbon basin compared with other margins of the Arabian plate. The horizontal extent of the Jurassic-Eocene shelf (i.e., post-Permian-Triassic neo-Tethyan rifting and separation and pre-Miocene Zagros collision and suturing) is taken from Hempton's (1987) tectonic reconstruction maps, which depict the northeast/north Arabian continental margin as having a very wide, stretched continental crust. Shortening and thrust stacking of this crust commenced in the late Eocene as continent-to-continent collision began, first with the Bitlis (Taurus) suture in the north and then with the Zagros suture along the northeast in the late Oligocene-early Miocene. The crust was restacked to its original thickness by the middle of the Miocene, and foreland folding along the whole margin began with some dextral horizontal motion taking place

Table 2—Source rocks of the Arabian plate (compiled from various sources).

INFRACAMBRIAN		YEMENS	JORDAN	S.E. TURKEY	SYRIA	N. IRAQ	S.W. IRAN	OFFSHORE IRAN	CENTRAL & S. IRAQ	KUWAIT & DIVIDED ZONE	SAUDI ARABIA	QATAR-BAHRAIN	UNITED ARAB EMIRATES	OMAN
TERTIARY	Neogene / Paleogene	Jezaa U. e Radhuma	Muwaqar	Germav	J a d d a l a	Aaliji	P a b d e h				Rudeis (Red Sea)			Andhur
CRETACEOUS	U / M / L	Mukalla	Ghareb Wadi Sir	Karabogaz Bakuk	Shiranish Soukhne	Balambo	G u r p i Khazhdumi Gadvan/Garau		Khasib Sulaiy Chia Gara/Yamama		Sulaiy	Khatiyah	Laffan Shilaif Shuaiba	Natih Kahmah
JURASSIC	U / M / L	Madbi/ Sabatayn				Sargelu	Sargelu		Najmah Sargelu		Hanifa/ Tuwaik Dhruma ? Marrat	Dukhan/Diyab Dhruma/Izhara		
TRIASSIC	U / M / L			Beduh	"Mulussa" (Kura Chine)				Baluti Beduh	Minjur?	? Jilh Sudair		Jilh	
PERMIAN	U / L													
CARBONIFEROUS					Markada	? Ora Khaista								
DEVONIAN											Jauf (Shales)			Misfar
SILURIAN	U / L		"Shales"	Bedinan	Tanf	? Chalki	Gahkum	? Gahkum	? Tabuk		Qusaybah	"Shales"		Safiq
ORDOVICIAN	U / L										Ra'an/ Hanadir			
CAMBRIAN														
INFRACAMBRIAN														Ara Buah Shuram

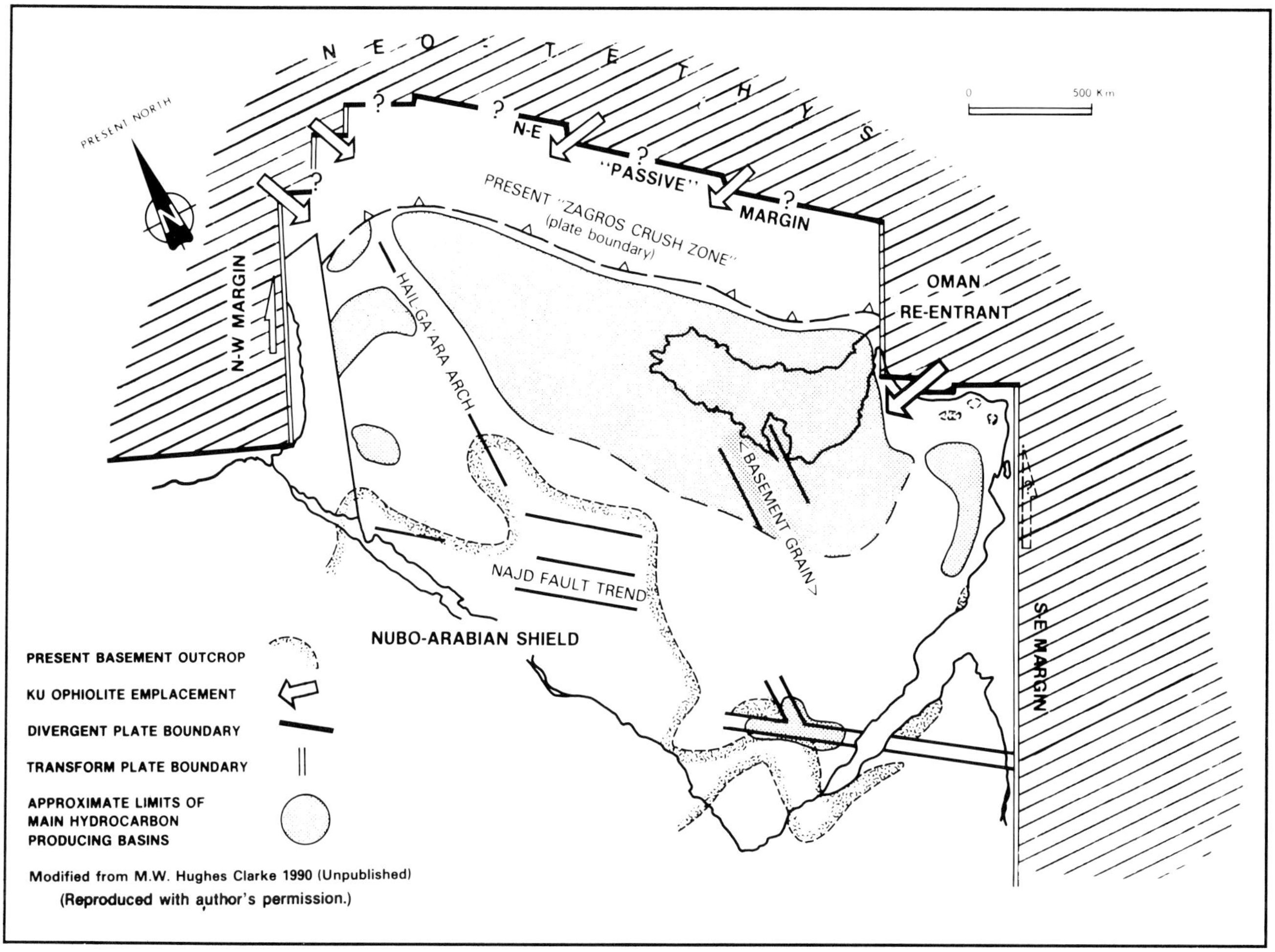

Figure 25—Conceptual sketch map of the "Arabian Promontory," an eastern spur of Africa.

along the Zagros suture. Folding and thrusting continue to the present (Hempton, 1987).

The regional Ha'il-Ga'ara (-Mardin/Buzova) arch forms the dividing line between this extraordinarily wide depositional northeast margin platform and the relatively narrow outer platform to its west, which had a much less stable history, particularly during the Cretaceous (Riche and Prestat, 1979). This probably accounts for the poor areal distribution of Cretaceous source rocks in the north and northwest margins during eustatic, transgressive sea level rises. There were also extensive Late Jurassic/Early Cretaceous block movements in the region accompanied by deep erosion of the elevated areas and blocks.

Murris (1980) developed a persuasive model for source rock-reservoir rock association over the extraordinarily wide northeast margin platform during the Mesozoic. Source rocks would have formed in starved basins during transgressive periods on the differentiated carbonate platform. Differentiated carbonate deposition develops sediment-starved intrashelf basins that are separated by high-energy margins from carbonate-evaporite shelves. Source rocks form in the shallow but stratified basins. Marginal mounds, oolite bars and shoals, rudistid banks and regressive sands form the main reservoirs, and supratidal evaporites and regressive shales form the seals. The spatial arrangement of these elements, together with adequate source maturity through burial and the early development and correct alignment of traps, account for the prolific Mesozoic hydrocarbon distribution.

Grunau (1977) further highlights this point with regard to the organic carbon content of all the Mesozoic source rocks of this region. He states that if the organic carbon content were only 5% and the total aggregate thickness of all potential source rocks did not exceed 200 m, still the magnitude of the area in which the source rocks underwent maturation is enough to account for the present hydrocarbon richness. Furthermore, traps were available to retain any oil being expelled from Paleozoic times onward as a result of the variety of trap-forming mechanisms available. Stoneley (1987, 1990b) pointed out that the organic matter of the Mesozoic potential source rocks of the region is almost entirely type II marine sapropel and that organic contents commonly exceed 10% and may reach 15%.

SOURCE ROCKS

The most important parameter for good hydrocarbon potential in any area is the availability of extensively deposited organic-rich source rocks. These must have matured sufficiently to be capable of generating hydrocarbons and charging suitably placed carrier formations in order to permit commercially viable amounts of hydrocarbons to accumulate in integral traps under efficient seals.

The Arabian plate region, especially its pre- and postrift northeast Gondwanan margin, is particularly blessed in the repeated wide distribution of potential source sediments throughout much of Phanerozoic time—although this deposition began in the late Proterozoic–Infracambrian. The source rock units for the Arabian plate region are listed in Table 2. A review of their main characteristics is essential as a basic step in assessing the petroleum potential of the entire region.

Infracambrian to Early Cambrian

The pre-Permian sequence along most of the Arabian plate region in general, and the northeast margin in particular, is deeply buried (Figure 8) and has seldom been penetrated by the drill. Hence, the distribution of Huqf-equivalent organic-rich source rocks of Infracambrian/Early Cambrian age, which source the oil in Paleozoic and younger reservoirs of the southeast Arabian margin, is presently only conjectural. By correlation of Huqf group deposits with other comparable successions elsewhere in the Arabian plate region and in then-adjacent portions of Gondwana (e.g. India, Central Iran), Gorin et al. (1982) compiled a paleogeography of the equivalents of the terminal Huqf carbonate-evaporite cycle and delineated the evaporite basins that presently survive in the Gulf–southwest Iran–Oman portion of the Gondwanan margin. Outside these basins they concluded that most of the remainder of the Arabian and adjacent region constituted carbonate–anhydrite platforms (Figure 7). Throughout the northeast margin, these deposits are found as manifestations of surface salt diapirs with entrained rock fragments over much of the Gulf region, and the original stratal position has rarely, if ever, been reached by the drill. Theoretically at least, the postulate that deeply buried Huqf-equivalent deposits, with or without evaporites, occur under most of the northeast margin enhances the hydrocarbon potential of the lower Paleozoic as well as the Infracambrian reservoirs of the whole region. The postulate also implies that similar accumulations to those that have been found along the southeast margin in Oman could be awaiting discovery under the thick, masking Phanerozoic cover. There are, however, some very critical questions that need to be addressed and that, although they cannot be answered adequately with present knowledge, diminish to some extent the theoretical prospects potential (Beydoun, 1989b).

The role of saline conditions may prove to have been all-important in the prospectivity of these deeply buried deposits. This would have applied in two principal ways. First, it promoted the generation of ample aqueous organic matter in the surface waters of the Infracambrian rift-related basins at a time when no terrestrial organic matter input was available. Second, it facilitated the accumulation of that organic matter because hypersaline density stratification was probably critical in allowing widespread anoxic bottom conditions to be maintained in these enclosed basins (Beydoun, 1989b). The role of evaporitic environments in enhancing source rock potential, with regard to organic production and preservation of organic matter, has been discussed by Kirkland and Evans (1981). From the available data on the areal distribution of the Ara and Hormuz evaporites, and by implication, their associated source rocks, it appears that the evaporites were restricted to the southern sector of the present northeast margin and to the southeast margin. These evaporites occur mainly in the Gulf area where frequently they are deeply buried and therefore hot and, today, probably well below the oil window with only dry gas potential left. Up dip, behind the wide frontal margin, it is possible that some intra-platform basins and grabens have developed evaporitic conditions that gave rise to more modest volumes of salts or sulphates that either are not recognizable on present subsurface data or subsequently were removed by dissolution. In such cases, their hidden or former presence will have to be recognized indirectly. Sharp structural disharmonies on seismic records of the overlying Paleozoic successions could be indicators of movement, dissolution, and removal of former deeper salt bodies (M.W. Hughes Clarke, personal communication, 1989). This aspect should be kept in mind in all areas within the Arabian plate.

The present great depth of the Infracambrian to Lower Cambrian source rocks in the Gulf region of the northeast margin need not detract greatly from their potential as initial generators of oils and gas now trapped in younger reservoirs. Grantham et al. (1988) noted the questionability of the assumption that the enormous reserves of oil in Jurassic reservoirs of the eastern Arabian region were entirely generated from the rich but thin Jurassic source rocks. They raised the possibility of additional input into this part of the basin from Huqf-related oil remigrating into younger reservoirs after earlier seals had become less effective or earlier traps had been displaced (e.g., by salt withdrawal and removal at salt edges within or at the margins of the basin). This is shown by Al-Marjeby and Nash (1986) to occur in southern Oman along the southeast margin. This additional input would have resulted in a history of multistage migration from the source through one or more paleotraps prior to the source becoming over-mature. This concept makes the less-deeply buried, older Paleozoic reservoirs updip along much of the northeast margin potentially prospective, provided an adequate definition of such paleotraps can be made in order to understand the geometry of prospect charge vectors. To recognize this, account must also be taken of possible changes in hydrocarbon characteristics during such long-term migration both in time and distance. Systematic geochemical analysis of all oils, matching them to source rocks and eventually, as techniques improve, matching gases to source rocks, is probably the method to achieve this level of understanding of the region's petroleum history and prospectivity.

Oils discovered in Oman have been geochemically typed and correlated with various source rock samples collected from Huqf group, as well as other potential source rock levels, such as the Safiq. The Huqf-generated oils are unusual and are characterized by the "X"-branched hydrocarbons. Huqf oils have a strong predominance of C_{29} or C_{27} steranes, a very low content of rearranged steranes, conspicuously negative carbon values of around –30‰ to –36‰, and a pristane/phytane ratio of less than 1 (Grantham et al., 1988; Sykes and Abu Risheh, 1989). The same geochemical features are also present in a tar obtained from an exploration borehole drilled in Pakistan. This was generated from Infracambrian source rocks much like those in the Huqf and the two basins were adjacent and possibly contiguous prior to the pre-Permian (Grantham et al., 1988).

Paleozoic Source Rocks

The sudden burst of organic diversity coupled with the richness of marine, and later of terrestrial, organic matter in the Paleozoic gave rise to several phases of source rock development. A number of the source rock phases have been identified within the Paleozoic sequences of the Arabian plate. They include locally significant horizons as well as regionally extensive ones linked to global sea level rise and development of anoxic conditions (Table 2). The most significant regional source-rock development phase was the extensive deposition of marine shales that occurred during transgressive cycles in the Late Ordovician and Early Silurian. The actual distribution (original deposition reduced by later erosion) of these so-called "Silurian" or "Gotlandian" source rocks is, unfortunately, not all that confidently understood on a regional scale, mainly because of the paucity of deep exploration wells down to that level. Also, the actual distribution of source rocks is conceptually quite variable because of the widespread erosion during the period from mid-Silurian to mid-Carboniferous when much of Arabia remained largely positive (Figure 10). Grunau (1985) conceptually assigned a prominent role as a kerogenous oil source to the Silurian shales on the eastern side of Arabia and perhaps over much of the Arabian Peninsula and Fertile Crescent region. Ala et al. (1980) on the basis of geochemical studies on the Iranian side of the Gulf, found that the Silurian Gahkum formation is in an advanced stage of maturity and is the most proximal source for the Upper Permian Khuff formation gas. Where this formation is less mature, this would also be the most proximal source for potential upper Paleozoic oil. At the Kuh-e-Gahkum exposures in southeast Fars province, southwest Iraq, near the Zagros suture, Bordenave and Burwood (1990) report over 100 m of black silty Silurian shale with residual TOC values varying from 1-4.3%. This is high for an overmature source rock that had reached the graphite stage. Reconstructions indicated, however, that these shales had been buried to at least 6000 m preceding the main Pliocene phase of the Zagros orogeny. In the western part of central Oman adjacent to the Rub al Khali basin, the Safiq formation of Late Ordovician–Early Silurian age has a high organic content in one of four cyclic shale members. This member has been identified as the source of the oil in the Haushi clastic reservoirs of the Sahmah and Hasira fields (Grantham et al., 1988) (Table 2).

The widely extensive carbonates of the Upper Permian Khuff formation, inter- bedded with or capped by anhydrites or shales, hold the enormous gas reserves of the countries of the Gulf region and central Oman. Until recently, no significant amounts of oil had been found in this reservoir although condensate is often found in association with the gas and some minor amounts of light oil had been reported. In 1986, however, commercial quantities of oil were discovered in the Yibal field of central Oman (Abu Risheh and Al-Hinai, 1989). Still more recently (1989–1990), significant amounts of light oil and gas were discovered in central Saudi Arabia at six different localities in the clastic deposits of the Permian-Carboniferous Unayza formation, which lies directly below the Khuff and is equivalent to the Haushi clastics of Oman (Table 2).

As discussed by Beydoun (1988, 1989b), the most proximal and thus most likely source for the oil and gas accumulations in the Permian-Carboniferous clastics and Permian carbonates of the Gulf region (including central Arabia) and west-central Oman, is the organic-rich Silurian shale. This is confirmed geochemically for the Sahmah and Hasira fields of western Oman by Grantham et al. (1988) and strongly indicated for southwest Iran Khuff gas generation (Ala et al., 1980). Also, El-Bishlawy (1985) and Ali and Sildawi (1989) suggested that Silurian shales were the source of the Khuff and pre-Khuff gas of the southern Gulf region, although indigenous sediments within the Khuff itself also were considered a possibility. This view was verbally endorsed by several local specialists during the OAPEC/ADNOC seminar on deep formations in the Arab countries held in Abu Dhabi in October 1989 (see Beydoun, 1989b; Loutfi and El-Bishlawy, 1989). If one accepts this view ascribing the principle source for Khuff and Haushi/Unayza (pre-Khuff) oil and gas accumulations mentioned above as being the Silurian shales, then the deeply buried Silurian kitchen of the Gulf area, and other parts of the Arabian plate where the Silurian is deeply buried, generated dry gas and lesser amounts of condensate. At the same time, in the peripheral, less deeply buried parts of the region such as central Saudi Arabia and west-central Oman (adjacent to the Rub al Khali basin), oil was available to charge the Khuff and pre-Khuff reservoirs. It will be

[3]McGillivray and Husseini (personal communication, 1991) confirm, on the basis of carbon isotope and biomarker data, that the Silurian Qusaiba shale is the principal source for the central Saudi Arabia light oil discoveries.

Table 3—Petroleum-producing formations of the Arabian plate.

Periods, Epochs, Ages (Modified form Dunnington 1967)		YEMEN	JORDAN	S.E. TURKEY	SYRIA	N. IRAQ	S.W. IRAN	OFFSHORE IRAN	CENTRAL & S. IRAQ	KUWAIT & DIVIDED ZONE	SAUDI ARABIA	QATAR-BAHRAIN	UNITED ARAB EMIRATES	OMAN
MIO-PLIOCENE	MIO-PLIOCENE M - U MIOCENE							Fars : G.						
OLIGO-MIOCENE	L-M MIOCENE OGLIO-				Jerribe : O.G. Chilou : O.G.	"Main" O.G. Limestone :	Asmari : O.G. (Kalhor)	Ghar/Asmari : G.O.	"Ghar" : O.		Belayim/Rudeis : G.O. (Red Sea basin)			
EOCENE/ PALAEOCENE	U - M EOCENE L.EOCENE-PALAEOCENE	Habshiya : O. (Gulf of Aden)		Sinan : O.	Jaddala : O.	Avana/Jaddala /Aaliji : O.		Jahrum : O.		Dammam : O.G. U. Radhuma : O.			Pabdeh : G.	
CRETACEOUS U	MAESTRICHTIAN U.CAMPANIAN L.CAMPAN-/SANTONIAN CONIACIAN			Garzan : O Beloka : O. Karabogaz : O.	Shiranish : O.G. Rmah : O.G. "Massive"	Shiranish/ "Pilsner" : O.	Ilam : O.	Gurpi : O.	Hartha : O.					
CRETACEOUS M	TURONIAN CENOMANIAN ALBIAN APTIAN		Ajlun Gp : O. (Hummar & & Na'ur)	Karababa : O. + Derdere : O. (Mardin)	Limestone : O. "Judea" : O.	Kometan/ Dokan / O. "Qamchuqa" :	Servak : O. Dariyan : G.	Magwa : O. (Mishrif) Ahmadi : O. Wara : O. Mauddud : O. Burgan : O.	Mishrif : O. Nahr Umr : O.	Wara : O. Mauddud : O. Burgan : O.	Mishrif : O. Rumaila : O. Ahmadi : O. Wara : O. Mauddud : O. Burgan : O. Shuaiba : O.	Ahmadi : O. Nahr Umr : O.	Mishrif : O. Mauddud : O. Thamama (Shuaiba) : O.	Wasia : O. G. Thamama (Shuaiba) : O.
CRETACEOUS L	BARREMIAN HAUTERIVIAN VALANGINIAN BERRIASIAN	Sabatayn : O, G.			Rutbah : O.	L. Qamchuqa and Gargu: O.	Fahliyan : O. G.	Fahliyan : O. (Yamama)	Zubair : O. Yamama : O.	Zubair : O. Minagish : O. (Ratawi)	Biyadh : (Zubair) : O. Bubwaib : O. Yamama : O. Sulaiy : O.		Thamama : O.	
JURASSIC U, M, L	PORTLANDIAN- M. KIMMERIDGIAN L. KIMMERIDGIAN OXFORDIAN/CALLOVIAN BATHONIAN BAJOCIAN LIASSIC L. LIASSIC	Amran : G.		Camurlu : G	Butmah : O. +			Arab: O. Arab D : O.	Najmah : O.	Marrat : O.	Hith : O. Arab A-C : O. Arab D : O Jubaila : O. Hanifa : O. Tuwaiq : O. Dhruma : O.	Arab A-C : G.O. Arab D / Darb : O. Araej : O.	Arab / Darb. O. Araej : O. Marrat : O.	
	TRIASSIC				Kurachine : O.G. (Mulussa) : G.O.	Kurachine : O.	Kangan : G.							
PALEOZOIC	PERMIAN PERMO-CARBONIFEROUS				Markadeh : G.		Dalan : G.	Dalan : G.		? Khuff : G.	Khuff : G. Unayza : O,G.	Khuff : G.	Khuff : G.	Khuff : O,G. Haushi : O.
	CARBONIFEROUS SILURIAN-DEVONIAN		"Sands" : G.	Katin : G.O.									Pre-Khuff : G.	Haima : O,G.
	CAMBRO-ORDOVICIAN													Haima : O. (Amin, Mahwis Ghudun)
	CAMBRIAN / INFRA-CAMBRIAN													Ara : O. Buah : G.

O = Oil G = Gas, as predominant production G after O = Gas alone in some fields Condensate in many of the non-Permian Gas fields most interesting to await the results of geochemical typing of oil/source for the recent discoveries in Central Saudi Arabia.[3] However, it should be kept in mind that in Oman, the Permian-Carboniferous Haushi clastic reservoir oil (with the exception of that of the Sahmah and Hasirah fields in westernmost Oman) and possibly the Khuff carbonate reservoir oil of Yibal, is principally from Huqf-related source sediments (Grantham et al., 1988; Abu Risheh and Al-Hinai, 1989).

Except in Yibal (west-central Oman), reported Paleozoic oil is restricted to the pre-Khuff clastics. The reason is unclear but may involve both seal integrity and migration routes. The intra-Khuff evaporites may be involved in this dichotomy, and also seem to have been responsible for the sour nature of the Khuff gas in hot/deep traps (Ali and Sildawi, 1989).

The implications of all the above on Paleozoic hydrocarbon potential of the Arabian plate, particularly updip in the northeast plate margin, are tremendous. Already, in southeast Turkey, the small discoveries of light oil and condensate in a Devonian Katin formation sandstone reservoir such as the Barbes deep well and others (Beydoun, 1988) and the oil-impregnated Lower Permian(?) or Devonian(?) Hazro sands exposed in the Hazro area, are considered to have been sourced from the Ordovician–Silurian organic-rich Bedinan formation shales (Table 2). Similar organic-rich Silurian shales seem the most probable source of the gas discovery in northeast Jordan (Risha basin) in an adjacent sandstone reservoir (Beydoun 1988, 1989b) because a large area of eastern Jordan has an extensive, thick subsurface development of organic-rich black shales of mostly Silurian age (NRA, 1986). These shales seem likely to have generated the small, light oil accumulation reported from the Wadi Sirhan basin of southeastern Jordan in underlying Ordovician sands. These Silurian shales are probably also responsible for the gas discoveries in a Permian-Carboniferous clastic reservoir recently reported from the southern Palmyrides intra-plate "aulacogen" of Syria, but a Carboniferous source (Markada) is also possible.

Other Paleozoic potential source rocks are of local significance or have not yet been shown to have sourced trapped hydrocarbons, or both. These include Lower Devonian shales in the Jauf formation of Saudi Arabia, the equivalent Misfar formation of Oman, Upper Devonian shales in southeast Turkey and Iraq, and probable Carboniferous source rocks in Syria (Table 2).

Mesozoic and Cenozoic Source Rocks

Triassic source rocks that have generated hydrocarbons appear to be confined to the edge of the northeast margin. Specifically, these source rocks are confined to the junction area of northeast Syria/northwest Iraq/southeast Turkey, and to the hinterland of the northern plate margin in the Palmyride aulacogen/basin. Elsewhere, in the Arabian Gulf region, Triassic source potential is only moderate. Loutfi and Abdel Sattar (1987) reported moderate source rock development in the Triassic formations of the Abu Dhabi region; sapropelic kerogen was the dominant organic matter and subordinate humic matter was also present. Pronounced gas shows were encountered in some of the offshore wells and the Jilh (Gulailah) formation appeared to be the most likely source since it exhibited every degree of maturity. The geochemical evidence from this region implies that considerable amounts of hydrocarbons have been generated from Triassic sources in this part of the margin. Hydrocarbons may be entrapped in as-yet-undiscovered Triassic reservoirs, and although the absence of regional seals (e.g., evaporites) lowers the hydrocarbon potential, more consistent exploration could be rewarding. Elsewhere, in northern Jordan and Lebanon and other parts of Iraq, similar source potential seems to exist. However, in other sectors of the Arabian plate, the Triassic facies appear unfavorable for organic matter preservation.

Most of the producing hydrocarbon reservoirs in the Arabian plate in general and in the northeast margin countries specifically, are Mesozoic in age (Table 3). Particularly, the producing reservoirs are Middle Jurassic to mid-Cretaceous, with subordinate Upper Cretaceous reservoirs in southwest Iran/Iraq and southeast Turkey and northeast Syria. Also, the number of Jurassic and Cretaceous source rocks, particularly in the northeast margin, is equally impressive (Table 2). There is no lack of confirmed candidates for having sourced the oils trapped in the Jurassic/Cretaceous reservoirs of the Arabian plate. Many source rocks are regionally extensive and have matured sufficiently to have generated oil trapped in these Mesozoic reservoirs. Geochemical studies have satisfactorily matched the Saudi Arabian Mesozoic oils analyzed to Mesozoic source rocks (Ayres et al., 1982), although it has been questioned whether the volume of mature source rocks is sufficient to generate the observed volumes of oil (Barker and Dickey, 1984). Moreover, researchers have shown that the vast reserves of oil in the Tertiary reservoirs of the Iraq-Iran Zagros fold belt remigrated from mid-Cretaceous reservoirs when Late Tertiary Zagros compressional folding breached the Cretaceous-Paleogene argillaceous-carbonate seals (Dunnington, 1967; Kent and Warman, 1972).

Elsewhere, in within-plate (or intra-plate) basins and along the south margin, Upper Jurassic and Cretaceous source rocks are well developed although areally limited (Table 2). Murris (1980), however, remarked that the Mesozoic source rocks of the eastern Arabian region are of good quality but are not unusually rich or thick compared with those in other basins elsewhere in the world. Grantham et al. (1988) took this constraint further and suggested, on the basis of the proven, rich Proterozoic/Paleozoic oil sources found in Oman, that the enormous amounts of oil found in the Mesozoic reservoirs of eastern Arabia may not have been sourced solely from the questionably adequate volumes of Jurassic (?) and Cretaceous source rocks present. Oil in eastern Arabia's Mesozoic reserves may have been augmented by additional input of Infracambrian/Early Cambrian Huqf-related oils to account for this imbalance. Because these deeper source horizons were probably through the oil window before the Mesozoic traps formed, retention in intermediate paleotraps followed by escape/remigration would be probable. In the Oman case, Sykes and Abu Risheh (1989) in fact estimated that about 80% of the oil found in the country, including the oil in Mesozoic reservoirs in the main fields of north-central Oman, are Huqf-related oils. A comparative analogy may be drawn by the charging of Lower Cretaceous Thamama reservoirs from Upper Jurassic Dukhan-Diyab generated source rocks in the U.A.E. in areas where the Late Jurassic Hith evaporitic seal is thin or absent (Hassan and Azer, 1985), again demonstrating the considerable degree of vertical migration possible. The postulated widespread distribution of Huqf-equivalent source rocks associated with evaporites under at least half of the eastern Arabian margin calls for serious examination of the likelihood of multistage migration. If the evaporites have been subsequently locally mobilized, withdrawn, or dissolved to be unrecognizable, the extent of the Huqf may even be greater. Such migration could be from extensive earlier source sediments (Huqf-related or Silurian) through one or several reservoirs and paleotraps to final entrapment. This concept makes the petroleum potential of the shallower, buried Paleozoic reservoirs updip and under good seal excitingly prospective. The fact that oils from the pre-Mesozoic sources are widespread and thus are worthy of directed exploration, would be proved best by regional geochemical studies of these oils, because the Oman oils seem to have very characteristic features that should be recognizable unless the features were lost during migration or by dilution.

Organic-rich source rocks of Late Cretaceous and Paleogene age are present in the northeast Arabian plate margin area, on the north and northwest margin regions and hinterlands, and in Yemen on the south margin (Table 2). They are, on the whole, immature except where they are locally deeply buried. This occurs mostly in intra-plate grabens in other parts of the Arabian region (e.g., Syria and western Iraq) but is also the case in the foredeep troughs of the extreme northern part of the northeast margin (e.g. northwest Iraq–northeast Syria/southeast Turkey), and there they have sourced the relatively small accumulations of that region (Table 3). However, the region is a meeting point of two differing continental margins with a complex tectonic history related to the ancient Ha'il-Ga'ara (-Mardin) arch, which forms the junction of the northeast and north margins. It also separates drastically differing shelf geography and geometry and subsequent tectonic uplift, fracturing, and deep erosion. The Paleogene oils of the Kuwait-Basra area and of northern offshore Iran are probably sourced from low-maturity indigenous sources (Dunnington, 1967), with the oils being rather heavy and sulphurous, since upward leakage from mid-Cretaceous reservoirs in this platform area is thought unlikely.

In the light of the evidence for upward migration into the Tertiary reservoirs presented by Dunnington (1967) and others (Kent and Warman, 1972; Ala et al. 1980), the voluminous oils in the Tertiary reservoirs of the northeast margin have been generated from Tertiary source rocks only locally and minimally. The limited occurrences of Tertiary oil in southeast Turkey and northwest Syria along the north margin appear to have been generated from low-maturity indigenous sources, while the hydrocarbons of the offshore regions of the south and southwest margins (the Gulf of Aden and the Red Sea) have been generated from rapidly matured, organic-rich Tertiary source rocks. The latter occurred as a consequence of the accelerated burial and higher heat flow associated with the Neogene rifting of these margins.

SEALS AND RESERVOIRS

The Arabian plate region does not lack potential reservoir formations. However, many such potential reservoirs are distant from seals and thus only act as migration paths. This particularly applies to the Paleozoic sandstone reservoirs of the region. Therefore, the most important controlling factor for hydrocarbon accumulations, second only to the presence of mature source rocks, is the presence of efficient seals in direct relationship with the reservoir formation as seal-reservoir pairs. The best producer reservoirs are those where rocks with good porosity and permeability lie directly beneath efficient, extensive seals, thus eliminating waste zones.

Table 3 shows all the currently producing reservoirs of the Arabian plate. All those in Paleozoic strata, with the exception of the local areally limited Infracambrian/Lower Cambrian Buah and Ara carbonates of the Huqf group of Oman and the extensive Upper Permian (to locally Lower Triassic) Khuff carbonates of the eastern Arabian region, are sandstone reservoirs sealed locally by shales of marine, deltaic, or floodplain origin. The Mesozoic and Cenozoic reservoirs of the Arabian plate are overwhelmingly carbonate, with subordinate Lower Cretaceous sandstone reservoirs in the Kuwait-Basra-upper Gulf area and of the Euphrates graben of Syria, the Oligocene-Miocene Ghar sands of the upper Gulf, and the Neogene sands of the Gulf of Aden and Red Sea offshore areas.

Many of the potential Paleozoic sandstone reservoirs, however, are distant from shale seals. Also, the development of reservoir-seal pairs within the Paleozoic clastic succession is, on the whole, areally limited in scale, having rapid lateral variations in "clean" reservoir intervals, which must reduce the effectiveness of specific reservoirs on a regional basis. Furthermore, the efficiency of the shale seals varies considerably with local conditions of thickness and intensity of tectonic disturbances or deformations. In cases where the seal was inefficient, upward migration of hydrocarbons would have continued until it was stopped by a competent seal, which may even include locally tight carbonates such as those that do occur in the Khuff formation. When these carbonates are porous and/or fractured and thus form good reservoirs, interbedded Permian-Triassic evaporites would form the barrier to further upward migration—as has happened in the case of the Huqf- derived oils of the southeast margin in Oman. Where evaporites are absent and interbedded shales are thin, potential reservoirs in good structural traps have frequently proved either barren or flushed as has been reported from the north and northwest parts of the Arabian plate.

For the Mesozoic reservoirs, seals are generally more numerous and also often are more effective regionally. In the Triassic of the northeast/north/? northwest margins, evaporite seals are localized and can be intermixed with argillaceous seals (red bed mudstones or shales). For the Upper Jurassic reservoirs, evaporites are the principal seals in the eastern Arabia–Gulf region and southwest Iran/northeast Iraq (Figures 12 and 13). These are mainly sabkha-type anhydrites but locally they also include salt. The age of the evaporites drifts diachronously downward as one moves north. Similar age evaporites with salt seal the littoral sandstone and carbonate Upper Jurassic reservoirs of the Marib-Shabwa region of Yemen. The Cretaceous reservoirs are essentially sealed by shales, which also double in some cases as source rocks. The seal efficiency varies considerably in the more tectonically deformed parts of the eastern Arabian plate and, in the Zagros fold belt region, has permitted upward leakage into younger Cretaceous and Tertiary reservoirs. These include Upper Cretaceous argillaceous carbonates, which act as seals in less deformed areas but were fractured sufficiently to receive oil accumulations and to be capable of oil production, albeit modestly—e.g., Ain Zalah field in northwest Iraq (Daniel, 1954). The oil movement from matrix pores into fractures in such reservoirs is slow and production leads to a rapid pressure drop and early depletion.

The Tertiary reservoirs of the eastern Arabian region (northeast and north margins) are sealed mainly by the regional lower-middle Miocene evaporitic seal (Lower Fars/Gachsaran and equivalents; Figure 14). This seal is extensively developed over most of the area, especially in the Zagros foothills belt. It provided the most effective uppermost seal for remigrated Cretaceous oil into Oligocene-Miocene Asmari reservoirs as the intensity of Zagros structural folding proceeded toward the rising fold front (Dunnington, 1967). Where the evaporitic cover was absent or breached in the eastern section of the foothills and fold belt, the volume of oil lost at the surface must have been prodigious. Kent and Warman (1972) suggested that numerous breached Asmari anticlines in the easternmost part may never have been oil fields because the Fars evaporites were removed or were absent and remigration led straight to the surface (Figure 16). The historic seeps that fueled the "eternal fires" of the Zagros belt document the validity of large-scale leakage. Other locally important Tertiary reservoirs in the northeast margin include Eocene-Paleocene carbonates of the Kuwait-Divided (Neutral) Zone area and of northwest Iraq that contain heavy, sulphurous oil. This oil may have been left en route during the charging of the overlying Asmari reservoir (Dunnington, 1967), although an indigenous immature source is a stronger possibility for the accumulations in the former area. The Fars gas accumulation in the northern Iranian offshore and perhaps the Kuwait area could be by gas leakage from a lower reservoir through tight intervening units that restrict oil migration. Along the south and southwest margins, the Tertiary reservoirs of the Gulf of Aden are sealed by anhydrite (Rus formation) and/or shale. The Neogene sandstone reservoirs of the Red Sea are sealed by shales in seal/reservoir pairs with the overall seal of the pre-evaporite sequences being the Miocene salt.

TRAPS

The structure-forming mechanisms that have operated in various parts of the Arabian plate in general and in the northeast margin in particular, where the prolific hydrocarbon accumulations occur, have operated singly or collectively since very early in the Phanerozoic. Traps were available to retain the earliest generation of hydrocarbons from the Huqf and related source sediments from early Paleozoic time onward. Although little is known of this early phase, except in Oman, an analogy can be drawn for other parts of the Arabian plate from the southeast margin region of Oman.

These structure-forming mechanisms include several types. (a) Deep-seated basement faulting occurred, which resulted in subtle closures by differential sediment compaction over basement fault blocks (horsts, tilted, and rotated fault blocks). This process included extension and rifting phases. The phases range in age from the late Proterozoic Najd faulting event through Permian-Triassic Gondwanan rifting and subsequent Jurassic-Cretaceous phases of Gondwanan breakup of parts of the Arabian plate, to Neogene rifting of the Gulf of Aden/Red Sea margins. (b) Salt-cored structures and/or salt (halokinesis) movements occurred, which resulted in continuing structuring as overburden load increased. Salt movement could have led to salt withdrawal and complex drape and facies/thickness changes in overlying beds. Progressive salt pillowing led to piercement in the deeper part of the Hormuz/Ara evaporite basin. Similar salt movements occurred in the Jurassic salt basin of Yemen and possibly Kuwait-Iraq, the Miocene salt basin of the Red Sea/Gulf of Suez and, to a limited extent, within salt beds of the Miocene evaporite basins of the northeast margin and the Triassic salt of the Palmyride basin. (c) Structures arose from straightforward basin loading with basin axis crowding between the main anticlinal uplifts. (d) Compressional folding towards the northeast and north margin occurred as a consequence of collisional events in the Late Cretaceous and the Late Tertiary Zagros-Taurus orogeny. Folding was additionally facilitated by decollement or disharmony in deep-seated Hormuz salt and in younger evaporites higher in the section (Figure 16). A further complication is the superimposition of the northwest-southeast Zagros style and the east-west Taurus style folds on the older basement grain-induced Arabian style folds (north-south). (e) Less significant areally, some transtensional/transpressional structures developed that were principally associated with strike-slip movements along earlier transform margins. These structures are now overprinted by later deformation. Those structures associated with late strike-slip movements along the Levant (Dead Sea) transform, the East Anatolian transform, and the Zagros suture have a horizontal motion component. Similar minor horizontal motions occurred elsewhere within the broad northeast, north, northwest and southeast margins. However, these motions are more evident in areas within the Arabian plate (Jordan, the Palmyrides and others), arising from young "slice" tectonics related to the rotation of the Arabian plate in the late Neogene.

The principal trap-forming mechanism associated with Huqf-generated oil in the southeast Arabian margin (Oman) is linked to Ara salt pillowing, withdrawal, and dissolution (Al Marjeby and Nash, 1986). The periodic removal of Ara salt (initially near the surface and later at depth) at the margin of the South Oman salt basin, and by analogy, elsewhere where Ara and Hormuz salt is present (Figure 1), played a dominant role in shaping the complex geology, and resulted in a number of major structural styles. Trap development has occurred from the Cambrian reactivation of basement faults to late Paleozoic uplift and erosion, and, during the Mesozoic movements with hydrodynamic regime activation. The main structural effects that developed include:

(a) collapse of masses of dolomite/shale formerly interbedded with the salt, thus forming residual mounds;

(b) contemporaneous salt movements that influenced the large-scale geometries of sedimentary bodies within the overlying Paleozoic succession;

(c) the distribution of Paleozoic sedimentary facies that was affected when the rates of removal either balanced or exceeded sedimentation;

(d) salt-induced subsidence that permitted preservation of parts of the succession otherwise eroded at the margin;

(e) salt movement that gave rise to folding and fracturing in the overlying Cambrian to Cenozoic sediments (Heward, 1990); and

(f) salt loss that led to structural inversion such that thick clastics already accumulated in rim synclines of growing pillows were inverted into "turtle backs."

Other salt-related structures include truncation traps on the flanks of the salt structures, as well as structural closures in anticlines above the cores of evaporitic residues and porous and fractured dolomites within such residues (Mattes and Conway Morris, 1990). Episodic down-dip salt removal along the flank of the main salt basin allowed Huqf oil to bypass the salt seal and to charge overlying reservoirs. By analogy with the above, similar effects would apply to salt movements in younger (Jurassic, Miocene) evaporitic units elsewhere in the stratigraphic sequence of the Arabian plate.

The main traps on the plains of the northeast Arabian margin are very large, gentle anticlines, most on north-south axes, which include quite a number of "super giant" and many "giant" oil fields, with smaller more subtle subsidiary closures found between them. The Zagros traps are mainly huge asymmetric northwest-southeast compressional whalebacks in successive rows of parallel elongated closures. These large and obvious structures have dominated as explorationists' targets. On the other hand, the gross areal extent and uniformity of reservoir facies makes the search for stratigraphic pinchout trends more difficult to locate and of less urgency compared with the still numerous available structural traps. The comparatively small areal extent of the surface and near-surface salt domes around the various Arabian margins initially made them of less interest compared with the giant anticlinal traps. Nevertheless, as the large anticlines are drilled, exploration in the northeast and southeast margins has moved to less evident structures. More subtle closures, salt piercement structures, and stratigraphic/structural combination traps are being tested and reveal new reserves in such areally limited and crowded areas as the Abu Dhabi region.

THE PROLIFIC NORTHEAST MARGIN: CONCEPTUALLY UNDEREXPLORED?

The phenomenal success of oil exploration along the northeast margin of the Arabian plate, whether measured in discovery rates, volumes of oil per meter drilled, development cost per incremental barrel/day, or other factors, is probably the main reason why so many of the later discoveries have not been developed (Beydoun and Dunnington, 1975; Beydoun, 1988). Nevertheless, because of the anticipated continually increasing world energy demands through and after the 1990s and a relative decline in hydrocarbon production from other regions, most countries of the northeast margin are seeking to greatly increase their sustainable oil production and export capacities. Moreover, in response to world concern over global warming and reduction in CO_2 emissions, the international demand for increased use of natural gas as a major source of energy will lead to increasing gas exports (LNG and by pipeline). Both of these objectives cannot be accommodated in the long term without continued, intensive exploration to establish the additional reserves needed to sustain the proposed increases in production.

Not many of the countries of the northeast margin—perhaps only Kuwait, Bahrain and Abu Dhabi—have reached the mature stage of exploration for the traditional reservoirs, which they have done by having drilled conventional Mesozoic levels in most or all of the mapped structural traps. Even in these countries, exploration for deeper reservoirs can lead to substantial increases in reserves, as was proven in Kuwait recently (Beydoun, 1988). Moreover, upward revision of reserve estimates, including earlier economically discounted/bypassed discoveries, is continuously taking place as better delineation of trap size, reservoir parameters and refinement of accepted recovery methods and factors occur in the region. Furthermore, artificial stimulation of fields in such mature areas by new processing/producing methods that previously were considered too expensive, coupled with wider applications of enhanced oil recovery methods, will add to ultimate recovery figures.

For most of the countries in the greater part of the northeast margin, however, exploration has not yet reached the mature stage. Exploration drilling of untested or newly identified structures continues, following detailed seismic mapping that involves continuously improving acquisition or reprocessing techniques. A good example is Iraq (Al-Khirsan and Al-Siddiki, 1989), where new fields not only have been discovered recently in the traditional Tertiary and Cretaceous reservoirs, but also in deeper/older Cretaceous and potentially even Jurassic reservoirs. The hydrocarbon potential of the Triassic reservoirs seems incompletely assessed in most of the region because most investigations of the Triassic have been "on the way down" to test gas objectives in the Permian Khuff. However, substantial volumes of oil may well have been generated regionally from Triassic source rocks as they seem to have been in Abu Dhabi (Loutfi and Abdel Sattar, 1987).

Probably the most important future exploration targets that will add major new reserves for the northeast margin region are the investigation of Paleozoic levels. Organic-rich Silurian source rocks are known to have generated major volumes of hydrocarbons in several parts of the region (see previous section titled *Source Rocks*). The recent, important light oil and gas discoveries made in central Saudi Arabia south of Riyadh in Permian-Carboniferous sand reservoirs are very likely sourced by Silurian shales.[4] Although both the Silurian source and the Paleozoic reservoirs it may have charged lie deep and through the oil window under most of the eastern part of the northeast margin, potential traps lie updip in west-central Saudi Arabia and western Iraq where the source rock subcrop distribution may be widespread (Enclosure 1 and Figure 10). Both the northwest and southeast margins of the Rub al Khali intraplate/epicratonic basin (Figure 1) are considered to have similar Paleozoic oil potential, but elsewhere in the Gulf region, only gas potential seems likely. Deeper Mesozoic reservoirs, including the Triassic in south Iraq through the Gulf to Oman, which could hold Paleozoic charge, are for the most part still within the oil window but at the deeper drilling limit. The Jurassic/Triassic potential of the Zagros fold belt in both southwest Iran and northeast Iraq is also similar, but effective drilling exploration requires careful positioning of test holes on the asymmetric anticlines.

Finally, lower Paleozoic hydrocarbon prospects for most of the updip sector of the southern (Gulf) region of the northeast margin, sourced by Huqf-related rocks generally associated with Hormuz evaporites, have yet to be explored. Exploration drilling locations must be specially targeted for this purpose and supported by intensive geochemical typing of oils. The prevalence of Huqf-related oils in Oman on the southeast margin in latest Proterozoic, Paleozoic, and Mesozoic reservoirs (Sykes and Abu Risheh, 1989) make such an investigation important and could revise major future exploration concepts in the region. Even that part of the northeast margin where no evidence of Hormuz evaporites is known, the northern sector, may prove more attractive for Huqf-sourced Paleozoic formations if evidence for some evaporite deposition and/or structuring can be established (see section titled *Source Rocks*).

As a postscript to help identify exploration targets in the northeast margin, delineation of younger intra-shelf-margin grabens where upper Mesozoic or even Tertiary source rocks may have been preserved and buried deep enough to accelerate maturation, could provide or enhance younger reservoir prospects not only in the northeast margin but elsewhere in other intra-plate basins. This proved the case in the Euphrates graben in Syria and the Marib-Jauf-Shabwa basins of Yemen. The so-called Anah graben, in the western Iraq portion of the northeast margin, which forms a link with the Euphrates graben across the northern Ha'il-Ga'ara (Khleisia) arch, is such a potential exploration area. There, Upper Cretaceous organic-rich source rocks are buried sufficiently deeply to have generated oil, as is evident in surface seepages and tar "lakes". Some of these source rocks may have charged older reservoirs across the bounding faults, as has happened in Syria. Other similar, young, fault-controlled, local basins must be present in different parts of the northeast margin region, but are concealed under very young molasse cover. Given sufficient sediment fill, they could have mature, organically rich source sediments of latest Cretaceous to Early Tertiary age that are known elsewhere but are immature in those locations.

THE SOUTHEAST MARGIN: TESTING GROUND FOR OLDER HYDROCARBONS

There are considerable volumes of hydrocarbons in place in the southeast Arabian margin shelf, principally in central and southern Oman-Dhofar (12B barrels alone in traps related to salt removal–Heward, 1990). These reservoirs have been abundantly sourced from Infracambrian/Early Cambrian Huqf group source rocks at least 550 Ma in age. However, geochemical work has demonstrated that some of these sources are

[4]McGillivray and Husseini (personal communication, 1991) confirm, on the basis of carbon isotope and biomarker data, that the Silurian Qusaiba shale is the principal source for the central Saudi Arabia light oil discoveries.

immature or barely mature, although the bulk are still in the oil window and first generated oil 400 Ma ago. The limited presence of gas in the southern Oman region may reflect leakage over this long time span (Sykes and Abu Risheh, 1989). The principal reason for this unusually long history of oil generation is the intermittent uplift of the southeast margin since the mid-Paleozoic, when it was still part of the northern Gondwanan passive margin. Since the Silurian, the whole of the southeast margin has been relatively positive (in contrast to the continuous subsidence of the northeast margin) and the top of the Huqf group lies at moderate to intermediate depths all along the interior portion of the southeastern coast. Huqf group vitrinite reflectance levels range from 0.60 (start of oil generation) to 2.4 (within the gas/condensate window) over most of the southeast margin (Sykes and Abu Risheh, 1989). In northern Oman and beneath the Oman Mountains foredeep, the Huqf group is now deeply buried but hydrocarbon generation even here may have continued until relatively recently when maximum depth of burial was reached and then exceeded. The Safiq ("Silurian") source rocks, however, are still at intermediate depth along the border of the Rub al Khali basin (Figure 1). Thus, neither the Ara/Huqf nor the Safiq sediments have been very deeply buried, and have had a long history of oil generation in many parts of the margin and hinterland (Mattes and Conway Morris, 1990).

In view of this long history of Huqf source rock hydrocarbon generation, the principal future hydrocarbon plays in the southeast margin will continue to be for Haima and Haushi clastic reservoir plays. There is an almost monthly drilling confirmation that many (small) untested traps associated with salt withdrawal and dissolution have reservoired commercial quantities of oil (Enclosure 1). Some of this oil is heavy and sulfurous, but this is considered to be due primarily to subsequent transformation (Al Marjeby and Nash, 1986), although some must undoubtedly be the result of low source rock maturity. Some of the south Oman fields such as Marmul contain vast reserves of this heavy oil; however, most is not recoverable without the application of EOR methods (Nandyal et al., 1983).

Additionally, in view of the presence of source rock potential in pre-salt Huqf group (Sykes and Abu Risheh, 1989), pre-salt as well as intra-salt Ara plays offer challenging objectives both within and outside the salt basins. However, it is uncertain whether the current absence of salt is indicative of nondeposition—together with the greater depth, this has detracted from the petroleum potential of such source/reservoir pairs even if shale seals are present. Alternatively, the salt may have been removed subsequent to deposition. Despite this, the Ara salt remains the best seal horizon to retain further charge, albeit gas-condensate accumulations.

A very challenging future exploration play would be the search for Silurian oil trapped in Haushi and Khuff reservoirs along the interior or Rub al Khali flank of the southeast margin in northern Dhofar and the adjacent part of Yemen. This play is especially interesting because the Silurian Safiq shales have sourced commercial oil accumulations at Sahmah and Hasirah fields in western Oman (Grantham et al, 1988).

Mesozoic objectives, whether from indigenous sources or from Huqf-related sources charging Jurassic and Lower to mid-Cretaceous carbonate reservoirs, or both, are confined to northern Oman because of the thin Mesozoic section in the south. Mesozoic traps may also be in the offshore regions of the southeast margin that are not yet adequately explored. Here, it is also possible that Tertiary source rocks are sufficiently mature locally to have started oil generation, in theory, but the geothermal gradient is normal to low, which weakens the merits of these exploration objectives.

The overthrust belt against the Oman mountain front in the interior parts of the margin offers Mesozoic prospects. However, the ruggedness of the terrain and acoustic complexity of thrusts limit cost-effective seismic acquisition and interpretation. Nonetheless, a number of encouraging boreholes have been drilled.

Finally, stratigraphic/permeability pinchout prospects have never been explored for seriously in the region in view of the abundance of structural traps to test. However, such prospects do offer subtle plays of basin flank accumulations in Paleozoic and some Mesozoic clastic pinchouts in the southern salt basins and the Rub al Khali basin.

THE NORTH AND NORTHWEST MARGINS: SMALL SO FAR, BUT...?

The principal oil-producing reservoirs of the north and northwest Arabian margins and of the intra-plate basins within them (southeast Turkey, Syria, Jordan) are of Cretaceous age, with subordinate oil, gas and condensate production from the Triassic strata of Syria (Palmyrides and north and northeast Syria and the Euphrates graben area). So far, there has been only minor hydrocarbon production from the Paleozoic (Table 3).

The proven recoverable hydrocarbon reserves of the north and northwest Arabian margins and their within-plate basins west of the Ha'il-Ga'ara arch are very small compared with the recoverable reserves of the northeast margin and relatively small compared with those of the southeast margin (Table 1). This mainly seems due to the absence of areally extensive mature source rocks as a consequence of the more complex genetic evolution of the region's tectonic components: dissection of the old Gondwana margin by the Paleozoic Ha'il-Ga'ara arch; Permian-Triassic rifting and basin inversion; middle Mesozoic block uplift and erosion; latest Cretaceous deformation, ophiolite obduction and further partial basin inversion; late Cenozoic frontal collision and thrusting and marginal horizontal motion and deformation. All these led both to differential uplift and erosion of much of the older succession throughout large areas, including the stripping of source rocks, or to differentiation of the platform by tectonic compartmentalization that resulted in far more areally limited and less uniform deposition of younger source rocks. Because many areas remained positive, or were elevated this also greatly reduced the width of the depositional platform, in comparison with that of the northeast margin, and limited the areas where these younger potential source rocks are mature. Thus, the old Paleozoic Gondwanan margin of this region was transformed into a mosaic of small basins and uplifts cumulatively overprinted by the successive episodes of fracturing and deformation. Thus the task of deciphering subcrops of source and reservoir/seal distribution under the Tertiary cover is very complex.

Until recently, the main oil production of the north margin (southeast Turkey and northeast Syria) has been principally from Upper Cretaceous reservoirs (Table 3) that were considered to have been charged from more basinal deeply buried Upper Cretaceous source rocks, such as the Soukhne, Shiranish, Karabogaz, and Germav (Table 2). Oil generation could not have been before about middle to late Tertiary times. The fact that most of the oils of this region are heavy and sulphurous is attributable both to the early stage of maturity of the source rock and to meteoric water flushing and biodegradation via outcrops of some of the reservoirs and/or insufficient sealing. The "Massive Limestone" reservoir of northeast Syria is considered to have been charged from Upper Cretaceous source rocks, but some additional charging via older reservoirs by upward leakage is possible (Metewalli, et al., 1974). For the middle Cretaceous Mardin group carbonate reservoirs of southeast Turkey, Granau (1977) favored a Paleozoic source such as the Bedinan shales. This is logical because in many parts of the southeast Turkey oil field area, the Cretaceous often lies unconformably on the Paleozoic. Ala and Moss (1979), however, preferred the Upper Cretaceous/lower Paleocene Germav shale as the most viable source for most of the oil of southeast Turkey.

In the last few years, a new play has been established in Syria, currently with at least 18 commercial oil discoveries of light-gravity, low-sulfur crude in Lower Cretaceous Rutbah formation sandstone reservoirs in the Euphrates graben (Enclosure 1). The source for this oil is now considered to be the widespread but generally immature, organic-rich Upper Cretaceous Soukhne/Shiranish argillaceous sediments. These sediments are sufficiently deeply buried in the graben to generate light oil that then charged older and younger reservoirs that were in suitable juxtaposition due to rift/block-faulting movements. Similar but younger, organically rich marls (Jaddala, Aaliji) have also sourced some minor accumulations in Syria, where they have been sufficiently buried locally (Metwalli et al, 1974; Ala and Moss, 1979).

Further back from the northwest margin in Jordan, the minor middle Cretaceous oil discovery in the Azraq basin is sourced by interbedded shales within the sequence, while the Silurian shales have sourced the gas accumulation in the Risha (Paleozoic) basin to the east (Beydoun, 1988). The source of the floating asphalt blocks in the Dead Sea and the oils and shows in the various wells drilled on the west flank of the Dead Sea pull apart basin, however, has been geochemically typed as the Upper Cretaceous (Qatrane; Ghareb) sediments (Spiro et al, 1983). These are organically rich argillaceous carbonates (equivalent to the Soukhne, etc.) that are buried sufficiently deeply in the Jordan structural depression for oil generation to have occurred.

The discovery, in Triassic carbonates, of important potential gas and condensate reserves in the Palmyride basin of the northwest margin with the possibility of oil accumulations down flank, as would be suggested from heavier (immature) small oil accumulations to the north of the basin at Sfayeh/Wahab (Beydoun, 1988), opens up very prospective plays in this region and in adjacent basins. This is particularly promising because small Triassic accumulations are already known from the Cretaceous oil field belt of northeast Syria and adjacent portions of Turkey/Iraq. Recently also, one or more Triassic carbonate reservoir discoveries were made in the Euphrates graben. These hydrocarbons appear to be sourced from indigenous shales within the Triassic carbonate-shale-evaporite cyclic sequence (Beydoun, 1988). Similar sediments were deposited over much of Syria, northern Jordan, and probably also under parts of Lebanon, most of which were not affected by later tectonic events and therefore offer viable exploration objectives.

Deeper light oil and gas accumulations in Paleozoic reservoirs have been found in scattered parts of the north and northwest margins, such as southeast Turkey, Syria, and Jordan. However, these discoveries have been minor so far. The presence of unevenly preserved but organic-rich, locally thick Silurian shales, opens up significant possibilities for Paleozoic discoveries as with the case of the hinterland of the northeast margin (Sections titled *The Prolific Northeast Margin* and *The Southeast Margin*). The subcrop distribution of these shales under a late Paleozoic unconformity, and the burial histories of the mainly inverted basins or the hidden grabens in which they accumulated, need to be better understood. Several either inverted Paleozoic basins or hidden grabens are present along the north and northwest margins and hinterlands from the Buzova-Mardin swell in the north to the Sirhan and Jafr basins in the south. Other inverted basins or grabens may be under the Neogene-Recent Hawran basalts of Syria-Jordan.

An attractive and as yet only sparsely tested play is the possibility of discovering commercial oil accumulations sourced by Upper Cretaceous organically rich, but normally immature, source rocks in pull-aparts along the current, new Levant (Dead Sea) northwest transform margin, where they would be buried sufficiently deeply to generate oil. Other than in the Dead Sea pull-apart, the Ghab basin to the north in Syria and possibly the Bekaa basin in Lebanon between these two, and the Aafrine and Kara Su basins of northwest

Syria/Turkey, all offer potential for commercial accumulations in reservoirs suitably juxtaposed with the deeply buried source rocks at the basin edge. Similar prospects occur in sub-thrust plays in the southern Palmyride basin edge, as well as in the deeper parts of similar rejuvenated basins in eastern Jordan.

In the Levant proper, which was the northwest margin until the Neogene, the best prospects may lie offshore. In the southern Levant, intensive exploratory drilling onshore has so far only produced modest results, largely because of limited Cretaceous source rock distribution and degree of maturity/charge vectors, and because thick, older Mesozoic carbonate reservoirs lack seals and lie at great depth. The northern Levant, by contrast, has been very poorly explored. Here, however, the pre-Jurassic is a challenging objective because of the likelihood of the extension of development of Upper or Middle Triassic evaporites linked to initial rifting episodes at moderate depths under the main Lebanon coastal uplift, as in the case of the adjacent Palmyride basin. Younger Mesozoic prospects in offshore Lebanon and Syria, charged and sealed by deeply buried Upper Cretaceous source rocks, may be more likely (Beydoun 1977, 1981). In the extreme north, there are high-risk offshore sub- (Upper Cretaceous) ophiolite (allochthonous) carbonate reservoir plays and Miocene reef prospects, both from indigenous source rocks in thick lateral basinal deposits in the Bay of Iskanderun region. Here, small amounts of Miocene oil have already been tested in reefal reservoirs (M. Rebora, personal communication, 1988).

THE SOUTH AND SOUTHWEST MARGINS: A YOUNG HISTORY THAT IS HYDROCARBON BEARING

The only current hydrocarbon production in the south and southwest margins of the Arabian plate is mainly oil but with major undeveloped gas reserves. In fact, this production is from an older, within-plate rift located inland from the junction area of the two margins, the Marib-Jauf-Shabwa (to Balhaf) graben (Enclosure 1). This Late Jurassic rift is filled with mainly Cretaceous sediments where it crosses the Neogene southern margin at the Gulf of Aden and extends to the Somali side of the Gulf (Beydoun, 1970). Oil production is from Upper Jurassic sand (Maycock, 1987, oral account) and carbonate reservoirs and averages about 200,000 bbl/day via a pipeline outlet on the Red Sea. A further outlet on the Gulf of Aden is near commissioning.

Elsewhere in this region, a series of exploration wells, many offshore, have been drilled both in the Gulf of Aden and the Red Sea (Figure 25). The results are one undeveloped oil discovery in an Eocene carbonate reservoir off the Yemeni Gulf of Aden coast at Sharmah, and another undeveloped gas, condensate, and light oil discovery in infra-evaporite Miocene sandstone reservoirs on the Arabian side of the northernmost sector of the Red Sea at Barquan. On the African side of the Red Sea basin, exploration drilling has led to two undeveloped post-evaporite Miocene sandstone reservoir discoveries (gas/condensate and gas, respectively) in offshore Sudan, and one unassessed gas blowout from an intra-evaporite Miocene carbonate stringer in offshore Ethiopia (Beydoun, 1989a; Bunter and Abdel Majid, 1989; Savoyat, et al., 1989; Haitham and Nani, 1990) (Table 3).

Prerift, organic-rich, bituminous source rocks have been identified in the Gulf of Aden from early work (Beydoun, 1964) as well as from the more extensive, recent geochemical work (Haitham and Nani, 1990; O'Connor, 1990) and are shown in Table 2. They include Upper Jurassic Madbi/Sabatayn marls and shales, Upper Cretaceous Mukalla and Sharwayn shales and marls, Paleocene Umm er Radhuma and Eocene Jezaa and Habshiya shales. TOC values in the Jezaa are about 1% but with a high hydrogen index (Haitham and Nani, 1990).

Equivalent Upper Jurassic source rocks are present only in the southernmost Red Sea, with the best quality being identified only on the African side (Savoyat, et al., 1989). Upper Cretaceous-Paleocene marine deposits are known from the northern half of the onshore Red Sea, in outcrops on the Arabian side and in Egypt and onshore wells in the Sudan. Some of these deposits could have source rock potential in the marl/shale zones if these extend offshore, which is debatable.

The synrift Oligocene-Miocene successions in the Gulf of Aden and Red Sea offshore basins include potential source rock shales in the Ghayda (Oligocene) and Hami and Sarar (Miocene and Pliocene) formations, which correspond to the infra-and post-evaporite levels of the Red Sea. However, these are mainly immature to marginally mature. The Gulf of Aden rift developed limited, restricted evaporitic facies in the Oligocene, but subsequently was much more marine, being more open to Indian Ocean circulation than was the Red Sea. In the Red Sea, potential source rock shales are present in both the infra-evaporite series (Rudeis or Kareem formation equivalents, which generated the hydrocarbons of the Barquan field of northernmost Saudi Arabia), and also in the supra-evaporite sequence that sourced the condensate and gas discoveries of the Sudan. Shales within the evaporite series are also oil prone (Bunter and Abdel Majid, 1989). The many oil seepages throughout the Red Sea basin, especially in the south (Figure 24), are evidence of black oil generation. Current geochemical and temperature gradient investigations (O'Connor, 1990) indicate that, contrary to conventional wisdom, the Red Sea is neither uniformly hot nor gas prone. The southern Red Sea, however, generally has a higher geothermal gradient than does the northern Red Sea.

The onshore area of the south margin possesses a number of pre-Neogene rift plays principally confined to Upper Jurassic objectives in those areas where Upper Jurassic evaporites also were deposited. This restricts the Jurassic play area to the region west of the Mukalla high (west of Long 49°E. on Figure 1) and perhaps the easternmost part of Yemen. In the western region, the structure is dominated by the main northwest-southeast-trending Marib-Jauf-Shabwa-Balhaf graben, which began to evolve and rapidly subside in the Kimmeridgian (Maycock, 1987, oral account). The graben contains considerable thicknesses of source, reservoir, and seal sediments far in excess of those developed on the adjacent shelf, and extensional block fault traps developed, totally hidden under flat, young alluvial cover. In addition to accumulations currently being produced, new discoveries have extended the prospect area in this main rift and offer similar prospects to the east. These prospects lie under flat Paleogene plateau cover, in subsidiary parallel rifts complicated by latest Jurassic/Early Cretaceous block uplift on north-south or north-northeast–south-southwest trends (e.g., the Mukalla high). What little published geophysical information there is, including aeromagnetic and gravity data and sparse regional seismic data, suggests the locus of an incipient "triple junction" in the Shabwa area that was subsequently overprinted and dissected by the terminal Jurassic block uplift. This triple junction appears to have evolved into a southeast-propagating failed rift whose southeast prolongation (Balhaf) subsided mainly in the Cretaceous, and an east-west-trending mainly Cretaceous failed rift (the Jezaa trough) to the east (Figure 1). A third structural depression with a thick sediment fill but with a trend that is difficult to determine is located to the northeast of the Shabwa area (the Sir basin). This depression may have a development of possible transitional evaporitic facies comparable to the Hith anhydrites of adjacent Saudi Arabia. There is no salt diapirism here, as in the adjacent Marib-Shabwa area (Paul, 1990). The petroleum potential of all these basins appears to be high because they are essentially comparable to the main Marib-Shabwa region in their development. The prospects in the Jezaa trough, however, also favor the Cretaceous sequence, with its thick marine facies that include potential source rocks, good reservoirs, and seals. The Jurassic appears present mainly in its eastern part.

In the hinterland of the south margin, the southern flank of the Rub al Khali basin lies adjacent to the Hadhramaut Arch and offers Paleozoic sandstone stratigraphic pinchout prospects against this arch. These pinchouts are of late Paleozoic age at least (Beydoun, 1988), and would be potentially sourced by mature Silurian shales in the deeper parts of the Rub al Khali basin. Younger prospects are less promising because the Jurassic is only thinly developed and the Cretaceous sequence on the main uplift of the arch is probably too shallow and distant from the oil kitchen in the north. However, sourcing of the Cretaceous from the Jazaa trough basin to the south is possible as demonstrated by the 7 bbl of 36° API oil produced in the Tarfayt well on the Hadhramaut arch (Paul, 1990). This same northern Hadhramaut arch region may contain some Infracambrian Huqf-type sediments preserved in an anomalous north-south-trending withinplate structural depression adjacent to Oman and indicated by gravity surveys (Beydoun, 1964). The thick section in this depression may also include a Jurassic sequence that outcrops on the coast to the southwest. In view of the total absence of the Jurassic over much of adjacent Dhofar (southwest Oman), however, such an assumption is speculative but does not detract from the petroleum prospectivity of this rift depression.

Within the Gulf of Aden rift, including the coastal plain, the petroleum prospects and plays are different. The presence of a potentially exploitable black oil accumulation at Sharmah (Figure 24) has been demonstrated. However, the field has not been developed because of the apparently relatively small size of the trap and the subsequently marginal economics for offshore development. The Arabian side of the Gulf of Aden rift has been sparsely explored by only 12 wells, 10 of them offshore (Figure 24), and most of those have been concentrated in the eastern part. Objectives here include prerift targets for Jurassic carbonate, Cretaceous carbonate (east) or sandstone (west) reservoirs, Paleogene carbonate, and Oligocene-Miocene sandstone reservoirs. The presence of about 2000 m of Oligocene (Ghayda formation) sediments in the area of the Sharmah discovery (Haitham and Nani, 1990) indicates that burial was rapid. This rapid burial accelerated the maturation of elsewhere immature or marginally mature source rocks such as the Paleogene Umm er Radhuma and Jezaa shales and the Shihr group shales and marls. Seals are provided by shales or by the Rus and Shihr group anhydrites.

The Paleogene deposits, however, do not extend very far west; the limit for the Umm er Radhuma carbonates, including the overlying Jezaa and equivalent Mejd Zir sands, is at about Long 46°E, while the shoreline for the Habshiya formation is at about Long 47°E. This makes the Aden/Abyan basin, with its rift-related volcanic activity, nonprospective for other than perhaps Jurassic plays. Jurassic plays would probably be reduced by intense magmatism as one approaches the Afar triple junction area of the western Gulf of Aden-southern Red Sea. Outside the influence of the Mukalla high, which lacks Jurassic strata, in the eastern part of the Gulf of Aden, the Jurassic sequence is at medium depths and within the oil window. However, near the mouth of the Gulf of Aden in the Qamar Bay/Jazaa trough syncline, the Cretaceous is thickly developed. This makes deep Jurassic prospects poor but Cretaceous prospects fair, especially because the shallow shelf here is very wide compared with elsewhere in the Gulf of Aden (Haitham and Nani, 1990).

The hydrocarbon prospects and plays of the southwest margin are entirely confined to the Red Sea rift basin because the uplifted onshore margin (rift shoulder) consists only of Precambrian metamorphic and igneous rocks. Targets are restricted to the syn- and postrift sequences, prerift sediments are believed to be absent within much of the Red Sea except in its southernmost part. Both pre- and post-evaporite plays offer potential as respectively demonstrated by the Barquan discovery on the Arabian side for the former, and the

Suakin and Beshayer discoveries in offshore Sudan for the latter. However, reliable pre-evaporite trap definition, suitable post-evaporite source rock maturity in adjacent sub-basins, and salt-related trap integrity remain as problems.

Tewfik and Ayyad (1982), using the currently available geothermal heat flow data, showed that the southern part of the Red Sea basin generally is hotter than the northern part. They suggested that oil generation in the south is at depths of 1000-2500 m, but for the north the range lies between 1500 and 4500 m. This would indicate that the Jurassic targets in the southernmost Red Sea, where very good source rocks are present (Savoyat et al., 1989), probably now lie past the oil and possibly even partially past the gas generation ranges. O'Connor (1990), however, reporting on current geothermal/geochemical data appraisal, modelling, and studies in progress for the UNDP/World Bank Red Sea-Gulf of Aden Regional Hydrocarbons Study Project (UNDP/RAB/88/O21), confirmed that the gradient slopes for the southern Red Sea (Ethiopian sector) are in general higher. He noted, however, that maximum recorded temperatures are similar to those in the northern Egyptian sector, due to the generally shallower (but often marginally so) drilling depths. Therefore, he questioned whether the southern sector is past the bottom range for oil generation. It should be kept in mind, in this connection, that high temperatures in the southern Red Sea are often local and due to the proximity of magmatic conduits such as those found in the Mansiyah well (Gillman, 1968).

The Red Sea basin, which constitutes the southwest Arabian plate margin, is thus an underexplored hydrocarbon-bearing basin with a total of only 51 exploration wells drilled. Eighteen wells are on the Arabian side, four of which were discoveries that remain undeveloped (Figure 24). Good pre- and post-evaporite Neogene shale source rocks and within- or intra-evaporite shales are present, which have generated the hydrocarbons in the four discoveries and in the oil seepages scattered throughout the basin (Beydoun, 1989a). The eastern or Arabian side is much less completely explored, although it probably possesses the better potential because it has a wider, shallow-water shelf and stretched continental crust basement with lower heat flow.

INTRA-PLATE BASINS: HARDLY EXPLORED; SOME MAY BE DEEP

The hydrocarbon potential of several of the intraplate basins of the Arabian plate has been mentioned in the context of the evolution of such basins in the framework of the respective adjacent plate margins. Amplification on the prospects of some of these is briefly repeated here, together with a brief review of the prospects of the remaining internal within-plate basins.

The Infracambrian/Early Cambrian sequence of the Jafr basin in southern Jordan has proved disappointing with regard to hydrocarbon generation from source rocks comparable to the Huqf of the southeast Arabian margin, although exploration drilling has been limited so far. Nevertheless, younger Paleozoic prospects sourced by mature Silurian shales, which are known to be present in the region, are still possible, provided the timing of trap formation was right. Similar prospects are possible in the adjacent but younger Sirhan/Tabuk basin in view of the gas discovery in the nearby Risha basin in northeast Jordan and the reported small oil occurrence in the Sirhan basin. Infracambrian basins related to the northwest-southeast Najd trend are present under the Rub el Khali and central Arabia portions of the Arabian plate, but the apparent absence of Infracambrian-Lower Cambrian evaporites, similar to the Ara-Hormuz, may detract from the prospects of such basins.

The inverted Paleozoic Risha basin of northeast Jordan has generally poor Mesozoic-Tertiary prospects except along the edge of what developed into the stable/unstable shelf boundary during the Mesozoic. This boundary forms the north and northwestern edges, where Triassic facies similar to those developed in the Palmyrides and northeast Syria are present. Traps here would be fault blocks or subtle, low-amplitude broad closures.

The Tabuk and Widyan basins of northwest Saudi Arabia/Jordan and northeast Saudi Arabia/southern Iraq include in their sequences Silurian and Devonian potential source rocks. In the Widyan basin, good Carboniferous/Permian Unayza sandstones unconformably overly subcrops of those potential source rocks, making late Paleozoic prospects as attractive as they have proved to be in central Saudi Arabia south of Riyadh (see section entitled *The Southeast Margin*). Additionally, Infracambrian J'balah group units have been noted in the Widyan (Al-Laboun, 1986). Although there is now no record of Lower Cambrian evaporites in the section, their former presence and subsequent removal are suggested by anomalous, high dips in the pre-Permian sequence of the basin in the proximity of the Ha'il (Jauf)-Ga'ara arch (Zamel, 1975). This has implications for the charging of younger reservoirs. The Mesozoic prospects of the Widyan basin are considered less attractive because potential source rocks are rarely sufficiently deep to have become mature.

The Rub al Khali basin has been relatively little explored. In the 1950s, a series of deep stratigraphic holes at strategic locations that were not on structural traps were drilled by Aramco to provide stratigraphic and velocity control for the seismic refraction survey over this vast Holocene dune-covered desert (Hoke and Zimerman, 1963). A number of wildcats were drilled in the eastern part, leading to several large but as yet undeveloped Mesozoic discoveries in the Saudi Arabian sector. The basin has a total thickness exceeding 8000 m, at least 2000 m of which are Paleozoic sediments (Figures 8 and 9). In the deeper parts, the Silurian organic-rich shales would now be in both the oil- and gas-generating phases. Light gravity oil, geochemically typed to these source rocks, is in Permian-Carboniferous Haushi group reservoirs of the Sahmah and Hasirah fields of western Oman on the southeast flank of the basin. The probability that these same source sediments have charged the equivalent age Unayza sandstone reservoir of central Saudi Arabia has already been discussed. The western part of the Rub al Khali basin, is almost totally unexplored but offers good prospects in older and younger Paleozoic reservoirs sourced by these Silurian shales. However, these shales undergo a facies change to coarser clastic facies in southwest Arabia, which affects source and seal and diminishes possibilities in Yemen (Figure 10). Infracambrian/Lower Cambrian Huqf type rocks are thought to underlie the Rub al Khali (Beydoun, 1989b) but in view of their great depth and the apparent absence of evaporites, their potential is regarded as questionable. The Jurassic succession, on the other hand, is well developed and resembles that of the prolific oil-producing area of eastern Arabia, including the presence of the terminal Arab-Hith anhydrite/carbonate cycles. Additionally, the marine Cretaceous oil-producing units of the Gulf region and their associated source rocks are well developed, making the Mesozoic prospects of this basin attractive. However, the logistics needed to retrieve any oil discovered in this hostile inland environment are daunting, particularly with the availability of other, more easily accessible discoveries still awaiting development.

The thick Cretaceous-Paleogene sequence preserved in the central Arabian graben and trough system offers prospects comparable to those of the similarly generated Euphrates graben system of the north margin. Maturation of potential pelagic shale beds within the otherwise predominantly carbonate reservoir facies of the Aruma group of northern Saudi Arabia, and the presence of overlying Umm er Radhuma shales in these structural basins, provide an interesting source/reservoir/seal association that should prove prospective. The central Arabian graben and trough system lies north of the part of central Arabia where the recent Permian-Carboniferous Unayza discoveries were made, and similar Paleozoic potential is also present there (Figure 10).

Chapter 10

Conclusions

The published recoverable petroleum reserves of the Arabian plate region at the end of 1990 were estimated at 663.2B bbl and 1,325.4 tcf, which represents almost 66.4% and 31.5% of the world's oil and gas reserves, respectively. Almost all of these reserves—98.7% of the oil and 97.2% of the gas—are located along the northeast Arabian plate margin in what has been referred to as the Iraq-Iran-Arabia basin, or the Gulf "Geosyncline." This area extends from northwest Iraq through the Gulf region to north-central Oman.

The remarkable concentration of prolific oil and gas fields along the northeast margin, many of which have multiple reservoirs and frequently are of giant or super-giant size, is intimately linked to the margin's long and stable history, essentially beginning in the earliest Phanerozoic. This stability permitted the evolution of a gently subsiding passive margin of unusual width that lasted from the late Paleozoic to the early Tertiary. On this virtually flat platform an almost uninterrupted accumulation of sheets of thick, uniform sediments were deposited. Differential sedimentation allowed areally extensive deposition of organic-rich source rocks in juxtaposition with equally extensive reservoir rocks and efficient seals. Furthermore, the continuous presence of large, often gentle structural traps, formed by a variety of mechanisms from early Paleozoic time onward allowed hydrocarbons to be trapped as source rocks matured and fluids were expelled. The margin's edge was modified into a collisional/compressional front in the middle Tertiary, and then was sutured with Eurasia. These events gave rise to additional structuring and accelerated lateral and vertical hydrocarbon migration and remigration (Enclosure 1). In contrast to this long history of stability, the other Arabian plate margins evolved differently.

There was a widespread deposition of Infracambrian/Early Cambrian sediments in fault-bounded basins, and Huqf group type sedimentary packages, associating source rocks with evaporites, probably were common over the paleo-Tethyan margin of Gondwana. The region that now forms the southeast Arabian margin experienced an initial phase of subsidence, but remained positive through the remaining Paleozoic. Gondwana segments broke away from Afro-Arabia beginning in the Permian-Triassic, but the southeast margin thus formed did not subside as a "normal" passive margin. The southeast margin's limited subsidence delayed maturation of the Infracambrian source rocks, allowing them to generate hydrocarbons up to the present. Charge from these old sources has fed traps in Paleozoic and Mesozoic reservoirs through complex migration to salt-based structures.

The north and northwest margins of the Arabian plate formed in the Triassic but later were subjected to substantial tectonic modification. Movements in the Late Jurassic/Early Cretaceous gave rise to dissection and block uplift with accompanying deep erosion. These processes occurred again in the Late Cretaceous when rifting, and later, ophiolite obduction, occurred and allochthons overran the platform edge. These events broke up the uniformity of the platforms on both margins and compartmentalized them, thereby greatly reducing the areal distribution of initially widespread source rocks and reservoir/seal pairs and giving rise to a complex subsurface mosaic of small basins and uplifts. From mid-Tertiary time, the northwest margin was rapidly replaced as the Arabian plate margin in favor of the new "Dead Sea" (Levant Fracture) transform margin to its east. At the same time, the north margin was modified and became collisional, also suturing to Eurasia. These events resulted in new, young structuration along the margin edges and gave rise to inversion of several interior within-plate basins.

The south and southwest margins formed in mid-Tertiary time as extensional rifts that developed into passive margins in the Miocene. These rifts developed when Arabia separated from Africa via propagation of the spreading Indian Ocean ridge through the Gulf of Aden to the Red Sea. The south margin, however, earlier had formed the distal interior shelf of the southeast Arabian-Somali margin of Afro-Arabia with similar Mesozoic marine deposits. A thick sequence of mainly Neogene sediments accumulated offshore in these rapidly subsiding rifts. These sediments are predominantly clastic but include evaporites that are most extensively developed in the Red Sea. Initially high heat flow favored the rapid maturation of the organic matter in these and other sediments.

The various within-plate basins studied throughout the Arabian plate have had differing evolutionary histories. These basin formation histories include initiation as late Proterozoic rifts, Paleozoic crustal sags with Mesozoic inversion, early Mesozoic rifting or Mesozoic rejuvenation of earlier fracturing, and Late Cretaceous-Paleogene rifts. Deciphering their subsurface source/reservoir/seal-combination distributions as subsurface data are acquired will undoubtedly answer the questions about their respective evolutionary histories and will lead to more confident assessment of the hydrocarbon potential of each basin. This approach has already succeeded in the south Oman salt basin, in the Euphrates graben, the Palmyride basin of Syria, and in the Marib-Shabwa basin of Yemen.

Recent geochemical typing of the oils produced from the Oman fields has shown that about 80% of the Oman oils are from Huqf group sources (Sykes and Abu Risheh, 1989). This includes not only those in the Paleozoic reservoirs of southern and central Oman, but also many of those from Jurassic and Cretaceous reservoirs in northern central Oman. Oil generation and migration has been going on since early in the Paleozoic, and when seals are inadequate or become breached, vertical migration into younger Paleozoic and Mesozoic reservoirs takes place. Since maturation and expulsion are still proceeding, even very recently formed traps can be charged when new hydrodynamic conditions are activated by new movements.

Huqf equivalent units in facies similar to those that have generated the Oman hydrocarbons, including the presence of evaporites, were deposited over most of the southern part of the northeast margin (the Gulf region) and probably also to a less-evident extent in the northern and hinterland parts of this margin. Thus, hydrocarbon generation conditions similar to those that have prevailed in Oman should also exist in those areas. The fact that Huqf beds on the northeast margin now lie buried deeply under the thick younger cover resulting from continued margin subsidence strongly suggests that these are now past the oil window and may even be locally through the gas window. This should not, however, detract from the likelihood of earlier-generated hydrocarbons charging younger reservoirs by remigration in response to intermittent regional movements, including salt flow and dissolution. Indicators of such progressive movements can be seen on the seismic record in a number of areas.

If hydrocarbons from older sources are indeed widespread, as the evidence in Oman strongly suggests, most Paleozoic (pre-Silurian, mainly sandstone) reservoirs in traps updip in the hinterland of the northeast margin are likely to contain hydrocarbons generated by Huqf type rocks. This Paleozoic sandstone belt has never been explored seriously in view of the great abundance of younger, and generally shallower, oil reserves in the mainly carbonate Mesozoic reservoirs of the margin. Thus high Paleozoic potential could exist all along the hinterland of the Saudi Arabian–Iraqi–sectors of this margin (Enclosure 1).

Silurian organic rich shales were deposited extensively over most of the Arabian plate region as a consequence of cyclical global sea level rises. The present subcrop distribution of these shales is not well defined because loss by subsequent erosion is not well revealed by the sparse deep well data at that level. However, Silurian sediments are inferred to have been preserved over extensive sectors of the region because they have generated hydrocarbons in western Oman where the oil has been geochemically typed, and also in southeast Turkey and Jordan where gas/condensate and light oil and gas, respectively, have been generated. Additionally, the consensus is that the Permian Khuff gas in the region originated in Silurian source rocks (Beydoun, 1989b). This is indirectly supported by geochemical data (e.g., Ala et al., 1980). Recently, oil has been found in the Khuff carbonate reservoir of Yibal in western Oman and although this oil may also be of Huqf origin, it also could have been generated by Silurian source sediments. The major light oil and gas discoveries made recently in central Arabia in the underlying Permian-Carboniferous Unayza sandstone reservoir, although not yet geochemically typed, are most plausibly suggestive of Silurian origin.[5] If Silurian sources are indeed widespread, gas would have been generated in the deeper parts of the basin and would have migrated up into the frequently evaporite-sealed and often tight Khuff carbonates. Oil would have been generated in the less deeply buried parts of the basin and would be trapped, generally at shale seals, in the underlying sandstones unless good permeability paths were present to permit migration to the Khuff carbonates such as may have happened in Yibal.

Proven sourcing from Paleozoic sediments makes all the updip hinterland areas of the northeast margin from Saudi Arabia to western Iraq and the peripheries of the Rub al Khali basin in western and southern Oman, Yemen, and central Arabia highly prospective for oil. Similar prospectiveness also would apply to the less deeply buried hinterlands of the north and northwest margins or to the flanks of the Mardin and Ha'il-Ga'ara arch, as well as to the inverted Paleozoic basins of Arabia. In the latter case, earlier depth of burial and degrees of maturation must be unravelled in order to establish the degree of prospectivity. This should not prove difficult to do as data from increased wildcat drilling for Paleozoic targets in Jordan and Syria is evaluated. Modest successes have been achieved in southeast Turkey and promising results are indicated in Jordan (Enclosure 1).

With the exception of areally small Bahrain and perhaps Kuwait and Abu Dhabi, no country in the prolific petroleum province of the northeast margin, let alone elsewhere in the Arabian plate region, has reached what may be termed the mature stage of petroleum exploration for the traditional reservoirs (Oligocene-Miocene, Lower and mid-Cretaceous, and Upper Jurassic). Even in these "mature" countries, exploration in deeper (Mesozoic and older) reservoirs can lead to substantial increases in reserves, as happened in Kuwait in the middle 1980s (Beydoun, 1988). Elsewhere, even the major producing countries, such as Saudi Arabia and Iraq, are sparsely explored. Considerable successes recently have been achieved in reserve enhancement in traditional reservoirs by the identification through widespread new geophysical

[5]McGillivray and Husseini (personal communication, 1991) confirm, on the basis of carbon isotope and biomarker data, that the Silurian Qusaiba shale is the principal source for the central Saudi Arabia light oil discoveries

surveys and test drilling of new, previously unrecognized structures (e.g., Iraq). Success has also come from the discovery of new Triassic, Jurassic, Cretaceous, and Tertiary reservoirs previously unrecognized or not considered as objectives (e.g., Iraq, Syria, southeast Turkey, and the Red Sea basin of Saudi Arabia). Finally, additional production has been established from traditional reservoirs in sectors or countries where there previously had been no production from these same reservoirs (e.g., parts of Syria, Yemen, and Jordan).

Upward revision of reserve estimates in the traditional reservoirs, because of better delineation of earlier, economically discounted or bypassed discoveries, takes place continuously with additional drilling and production experience and economic revision. Additionally, artificial stimulation of fields using methods previously considered too expensive, or application of EOR methods to stimulate declining production in long-established fields or both, will add considerably to the published reserve figures.

Accurate estimates of as-yet-undrilled oil and gas reserves in the region cannot be assessed with any high degree of probability, because we lack sufficient published data. In any case, the immaturity of exploration means that the new data that are continuously acquired markedly revise any current estimates. This occurs even where these necessary data have been made available, for such parameters as trap size, time of hydrocarbon generation, storage capacity, and seal efficiency, thus permitting one to calculate the size of the source drainage areas, thickness, and areal extent of source rocks, and zones of oil and gas maturity. However, using gross assumptions some volumetric calculations have been attempted for specific source/reservoir relationships in specific areas or subbasins. Barker and Dickey (1984) made such an estimate for the Jurassic of eastern Arabia. Their estimate resulted in questioning the ability of Jurassic source rocks alone to account for the oils trapped in Jurassic reservoirs of that area, and was an early indicator of the possibility that older source rocks contributed input into these reservoirs. Similar arguments were made by Dunnington (1967) and others to infer that the oil reserves in the Tertiary reservoirs of the Zagros fold belt were derived by vertical migration and remigration from Cretaceous-generated oils. More recently, similar arguments suggest that the oils of Oman in Paleozoic and some Mesozoic reservoirs originated from Huqf type source rocks (Grantham et al., 1988; Sykes and Abu Risheh, 1989). Notwithstanding this, accurate, sophisticated estimates for the whole region cannot be made.

In summary, estimates of ultimate oil and gas reserves for the region must be subjective and will vary considerably. With the benefit of hindsight, it is obvious that all Middle East regional reserve estimates and all revisions of earlier estimates made thus far have been extremely cautious, conservative, and hence, are grossly understated. Since Zarrug and Bois (1983) estimated the future additional recoverable oil reserves for the Arab countries of the Arabian plate (excluding Iran) to be between 60 and 220B bbl (Arabian plate published reserves at the end of 1982, including Iran were 369B bbl), the upper limit has already been exceeded by 36B bbl without taking into account Paleozoic reserves. Similarly, Grunau (1985) estimated the undrilled future oil potential for the same countries as roughly 150-200B bbl and the gas potential to be between 400 and 550 tcf. Those maxima have already been exceeded (by 51B bbl of oil) or are close to achievement (short by 121 tcf of gas), respectively. Today, the Paleozoic oil potential of the region, outside Oman, is virtually untouched and unassessed. However, the Paleozoic is beginning to indicate tremendous promise both as a consequence of establishing the presence of Huqf-type, Silurian, and possibly other Paleozoic sources and especially by the continuing major discoveries in central Arabia. Current reserve figures could easily increase by half or, conceivably, even double in another decade or two of intensive exploration. Clearly, the Arabian plate region will dominate world hydrocarbon reserves well after other currently producing major hydrocarbon regions have declined to insignificance.

ACKNOWLEDGMENTS

This volume has greatly benefited from detailed reading and discussion by Dr. M.W. Hughes Clarke and by a critical review by Professor J.F. Dewey, F.R.S. Their input led to many valuable suggestions for improvements and balance; for these the author is extremely grateful. Any remaining omissions and shortcomings are thus entirely the author's responsibility.

Much gratitude is expressed to the International Exploration Management of Marathon Oil Company in Houston and the Exploration Management of Marathon International Petroleum (G.B.) Ltd. in London, for the considerable support provided both in time and facilities during the preparation of this review and for encouragement to publish it. Sincere thanks are due to the Marathon Drawing Office, London, for preparation of the figures, and to Miss Susan Hambrook and Miss Elizabeth Jenyon for patiently typing several corrected versions of the manuscript.

References Cited

ABU RISHEH, A. and Al-HINAI, K.B.M., 1979. Oil prospectivity of the Khuff formation in Oman. Proc. 6th SPE Middle East Oil Technic. Conf., March 1989, Bahrain, SPE 18008, 811-818.

AHMED, S.S., 1972. Geology and petroleum prospects in the eastern Red Sea. AAPG Bulletin 56:707-719.

ALA, M.A. and MOSS, B.J., 1979. Comparative petroleum geology of SE Turkey and NE Syria. Journ. Petrol. Geol., 1(4): 3-27.

ALA, M.A., KINGHORN, R.R.F. and RAHMAN, M., 1980. Organic geochemistry and source-rock characteristics of the Zagros petroleum province, SW Iran. Journ. Petrol. Geol., 3:61-89.

ALI, A.R. and SILDAWI, S.J., 1989. Hydrocarbon potential of Paleozoic pre-Khuff clastics in Abu Dhabi, U.A.E. Proc. 6th SPE Middle East Oil Technic. Conf., March 1989, Bahrain, SPE 18009, 819-832.

ALAVI, M., 1980. Tectonostratigraphic evolution of the Zagrosides in Iran. Geology, 8:144-149.

AZZAROLI, A., 1957. Missione Geologica in Migiurtina. La Ricerca Scientifica, 27(2):301-346.

AZZAROLI, A., 1968. On the evolution of the Gulf of Aden. 23rd Internat. Geol. Cong. Report, Prague, 1:25-34.

AZZAROLI, A., and FOIS, V., 1964. Geological outlines of the northern end of the Horn of Africa. 22nd Internat. Geol. Cong. Report, New Delhi, 4:292-314.

BACH-IMAM,. I. and SIGAL, J., 1985. Precisions nouvelles sur l'age triassique, et non-jurassique, de la majeur partie des formations evaporitques et dolomitiques des forages de l'est syrien. Rev. Paleobiologie, 4: 35-42.

BAKER, N.E. and HENSON, F.R.S., 1952. Geological conditions of oil occurrences in the Middle East fields. AAPG Bulletin, 36: 1885-1901.

BARAKAT, H. and MILLER, P., 1984. Geology and petroleum exploration, Safaga concession, northern Red Sea, Egypt. 7th EGPC Exploration Seminar, Cairo, 191-198.

BARKER, C. and DICKEY, P.E., 1984. Hydrocarbon habitat in main producing areas, Saudi Arabia: discussion. AAPG Bulletin, 68 : 108-109.

BARNES, S.U., 1976. Geology and oil prospects of Somalia, East Africa. AAPG Bulletin, 6O-389-413.

BEIN, A and GVIRTZMAN, G., 1977. A Mesozoic fossil edge of the Arabian plate along the Levant coastline and its bearing on the evolution of the eastern Mediterranean. In : Proc. Internat. Sympos. on the Structural History of the Mediteranean Basins, Split, B.Biju-Duval and I. Montadert, edits., Editions Technip, Paris, 95-110.

BENDER, F., 1974. Geology of Jordan. Borntraeger, Berlin, Stuttgart, 196p.

BERBERIAN, M. and KING, G.C.P., 1981. Towrds a paleogeography and tectonic evolution of Iran. Canad. Journ. Earth Sci., 18:210-265.

BERHE, S.M., 1986. Geomorphic and geochronologic constraints on the evolution of the Red Sea–Gulf of Aden and Afar depression. Journ. African Earth Sciences, 5:101-117.

BEYDOUN, Z.R., 1964. The Stratigraphy and Strucutre of the Eastern Aden Protectorate. Overseas Geology Mineral. Resour. Bull Supp. 5, H.M.S.O. London, 107p.

BEYDOUN, Z.R., 1970. Southern Arabia and northern Somalia : comparative geology. Phil. Trans. Roy. Soc. London, A267 : 267-292.

BEYDOUN, Z.R., 1977. Petroleum prospects of Lebanon : re-evaluation. AAPG Bulletin, 61: 43-64.

BEYDOUN, Z.R., 1981. Some open questions relating to petroleum prospects in Lebanon. Journ. Petrol. Geol., 3 : 304-314.

BEYDOUN, Z.R., 1982. The Gulf of Aden and Northwest Arabian Sea. In: The Ocean Basins and Margins, 6, The Indian Ocean, A.E.M. Nairn and F.G. Stehli, edits., Plenum Publish. Corp. New York, Ch. 6:253—313.

BEYDOUN Z.R., 1986. The petroleum resources of the Middle East : a review. Journ. Petrol. Geol., 9 : 5-28.

BEYDOUN, Z.R., 1988. The Middle East: Regional Geology and Petroleum Resources. Scientific Press, U.K., 292p.

BEYDOUN, Z.R., 1989a. Hydrocarbon prospects of the Red Sea–Gulf of Aden : a review. Journ. Petrol. Geol., 12 : 125-144.

BEYDOUN, Z.R., 1989b. Hydrocarbon potential of the deep (pre-Mesozoic) formations in the Middle East Arab countries. In:OAPEC/ADNOC Seminar on Deep Formations in the Arab Countries: Hydrocarbon Potential and Exploration Techniques, Abu Dhabi, Oct. 1989, Proc. Techncial Papers: 31-84.

BEYDOUN, Z.R. and GREENWOOD, J.E.G.W., 1968. Aden Protectorate and Dhufar. In: Lexique Stratigraphique International, Vol. III, Asie, L. Dubertret, edit., CNRS, Paris, fasc. 10 b2, 128p.

BEYDOUN, Z.R. and DUNNINGTON, H.V., 1975. The Petroleum Geology and Resources of the Middle East. Scientific Press, U.K., 99p.

BISHLAWY, S.H. El-, 1985. Geology and hydrocarbon occurrences of the Khuff formation in Abu Dhabi, U.A.E. Proc. 4th SPE Middle East Oil Technic. Conf., March 1985, Bahrain, SPE 13678, 9-24.

BLENDINGER, W., van VLIET, A., and HUGHES CLARKE, M.W., 1990. Updoming, rifting and continental margin development during the Late Paleozoic in northern Oman. In: The Geology and Tectonics of the Oman Region, A.H.F. Robertson, M.P. Searle, and A.C. Ries, edits. Geol. Soc. Special Publication 49 : 27-37.

BOIS, C.. BOUCHE, P. and PELET, R., 1982. Global geologic history and distribution of hydrocarbon reserves. AAPG Bulletin, 66:1248-1270.

BONATTI, E., 1985. Punctiform initiation of seafloor spreading in the Red Sea during transition from a continental to an oceanic rift. Nature, 316:33-37.

BORDENAVE, M. L. and BURWOOD, R., 1990. Source rock distribution and maturation in the Zagros Orogenic Belt: provenance of the Asmari and Bangestan reservoir oil accumulation. Org. Geochem., 16:369-387.

BRAAKMAN, J.H., LEVELL, B.K., MARTIN, J.H. POTTER, T.L. and Van VLIET, A., 1982. Late Palaeozoic Gondwana glaciation in Oman. Nature, 299:48-50.

BROWN, G.F. SCHMIDT, D.L. and HUFFMAN, A.C. Jr., 1989. Shield Area of Western Saudi Arabia. In: Geology of the Arabian Peninsula. Geol. Surv. Profess. Paper 560-A, U.S. Geol. Surv., 188p.

BUNTER, M.A.G. and ABDEL MAGID, A.E.M., 1989. The Sudanese Red Sea : 2. New development in petroleum geochemistry. Journ. Petrol. Geol. 12:167-186.

CARELLA, R. and SCARPA, M., 1962. Geologic results of exploration in Sudan by AGIP Mineraria. 4th Arab Petroleum Cong. Proc., Beirut, 27(B-3), 23p.

CGMM (Commission for the Geological Map of the World), 1976. Geological World Atlas, scale 1 : 10,000,000, sheet 8, UNESCO, Paris.

COCHRAN, J.R., 1981. The Gulf of Aden : structure and evolution of a young ocean basin and continental margin. Journ. Geophys. Res. 86, B1 : 263-287.

COCHRAN, J.R., 1983. A model for development of Red Sea. AAPG Bulletin, 67: 41-69.

COLEMAN. R.G., 1981. Tectonic setting for ophiolite obduction in Oman. Journ. Geophys. Research, 86(B4): 2497-2508.

COLEMAN, R.G., 1984. The Red Sea : a small ocean basin formed by continental extension and sea floor spreading. 27th Internat. Geol. Cong. Report, Moscow, 23 : 93-121.

COURTILLOT, V., GALDEANO, A. and Le MOUEL, J.C., 1980. Propagation of an accreting plate boundary : a discussion of new aeromagnetic data in the Gulf of Tadjurah and southern Afar. Earth Planet. Sci. Letts., 47 : 144-160.

DANIEL, E.J., 1954. Fractured resservoirs of the Middle East. AAPG Bulletin, 38 : 774-815.

DELFOUR, J., 1970. La groupe de J'Balah, une nouvelle unite du Bouclier arabe. B.R.G.M. Bulletin (2e Ser.), IV(4) : 19-32.

DERCOURT, J., ZONENSHAIN, C.P., RICOU, L-E, KAZMIN, V.G., LE PICHON, X., KNIPPPER, A.L., GRANDJAQUET, C., SBURTSHIKOV, I.M., GEYSSANT, J., LEPVRIER, C., PECHERSKY, D.H., BOULIN, J., SIBUET, J-C., SAVOSTIN, L.A., SOROKHTIN, O., WESTPHAL, M., BAZHENOV, M.L., LAUER, J.P., and BIJU-DUVAL, B., 1986. Geologic evolution of the Tethys belt from the Atlantic to the Pamirs since the Lias. Tectonophysics, 123 : 241-315.

DEWEY, J.F., PITMAN, W.C., RYAN, W.B.F.; and BONIN, J., 1973. Plate tectonics and the evolution of the Alpine system. Geol. Soc. America Bulletin, 84 : 3137-3180.

DEWEY, J.F., HEMPTON, M.R., KIDD, W.S.F., SAROGLU, F., and SENGOR, A.M.C., 1986. Shortening of continental lithosphere : the neotectonics of eastern Anatolia–a young collision zone. In: Collision Tectonics, M.P. Coward and A.C. Ries, edits., Geol. Soc. Special Publication 19 : 3-36.

DOW, D.B., BEYTH, M. and HAILU, T., 1971. Paleozoic glacial rocks recently discovered in northern Ethiopia. Geol. Mag. 108 : 53-60.

DUNNINGTON, H.V., 1958. Generation, migration, accumulation and dissipation of oil in northern Iraq. In: Habitat of Oil, AAPG Symposium, L.G. Weeks, edit., 1194-1251.

DUNNINGTON, H.V., 1960. Some problems of stratigraphy, structure and oil migration affecting the search for oil in Iraq. 2nd Arab. Petrol. Cong. Proc., Beirut, 18(B-2), 25p.

DUNNINGTON, H.V., 1967. Stratigraphical distribution of oilfields in the Iran–Iraq–Arabia basin. Journ. Inst. Petrol., 53 : 129-161.

DUNNINGTON, H.V., 1974. Aspects of Middle East oil geology. In: Geological Principles of World Oil Occurrences Proc., Banff, Alberta, 89-156.

DUNNINGTON, H.V., Van BELLEN, R.C., WETZEL, R. and MORTON, D.M., 1959. Iraq. In: Lexique Stratigraphique International, Vol III, Asie, L. Dubertret, edit., CNRS Paris, fasc 10a, 333p.

GASS, I.G. and MASSON-SMITH, D., 1963. The geology and gravity anomalies of the Troodos massif, Cyprus. Phil. Trans. Roy. Soc. London, A255:417-467.

GEALEY, W.K., 1977. Ophiolite obduction and geological evolution of the Oman Mountains and adjacent areas. Geol. Soc. America Bulletin, 88:1183-1191.

GETTINGS, M.E., BLANK, H.R., MOONEY, W.D. and HEALEY, J.H., 1986. Crustal structure of southernwestern Saudi Arabia. Journ. Geophys. Research, 91, B6 : 6491-6512.

GEUKENS, F., 1960. Contribnution a la geologie du Yemen. Inst. Geol. Louvain, Mem., 21 : 122-179.

GEUKENS, F., 1966. Yemen. In: Geology of the Arabian Peninsula. Geol. Surv. Profess. Paper 560-B, U.S. Geol. Survey, 23p.

GILLMAN, M., 1968. Primary results of geological and geophysical reconaissance of the Jizan coastal plain in Saudi Arabia. Proc. 2nd SPE Regional Technic. Sympos., Dhahran, Saudi Arabia, 189-212.

GILLMAN, M., LETULLIER, A. and RENOUARD,.G., 1966. La Mer Rouge, geologie et probleme petrolier. Rev. Inst. Francais du Petrole, 21 : 1467-1487.

GIRDLER, R.W., 1985. Problems concerning the evolution of oceanic lithosphere in the southern Red Sea. Tectonophysics, 116 : 109-122.

GIRDLER, R.W. and STYLES, P., 1974. Two stage sea floor spreading. Nature, 247 :7-11.

GIRDLER, R.W. BROWN, C., NOY, D.J.M. and STYLES, P., 1980. A geophysical survey of the westernmost Gulf of Aden. Phil. Trans. Roy. Soc. London, A298 : 1-43.

GIRDLER, R.W. and UNDERWOOD, 1985. The evolution of early oceanic lithosphere in the southern Red Sea. Tectonophysics,. 116 : 95-108.

GIRDLER, R.W. and SOUTHERN, T.C., 1987. Structure and evolution of the northern Red Sea. Nature. 330 : 716-721.

GLENNIE, K.W., 1977. Outline of the geology of Oman. Mem. h. serie., Soc. Geol. France, 8 : 25-31.

GLENNIE, K.W., BOEFF, M.G.A., HUGHES CLARKE M.W., MOODY-STUART, M., PILAAR, W.F.H. and REINHARDT, B.M., 1973. Late Cretaceous nappes in Oman Mountains and their geologic evolution. AAPG Bulletin, 57 : 5-27.

GLENNIE, K.W., BOEFF, M.G.A., HUGHES CLARKE, M.W., MOODY-STUART, M., PILAAR, W.F.A. and REINHARDT, B.M., 1974. Geology of the Oman Mountains. Konikl. Nederlands Geol. Mijnbowkundig Genoots. Verh., 31, 423p.

GLENNIE, K.W., HUGHES CLARK, M.W., BOEFF, M.G.A. PILAAR, W.F.H. and REINHARDT, B.M., 1990. Inter-relationship of Makran-Oman Mountains belts of convergence. In: The Geology and Tectonics of the Oman Region, A.H.F. Robertson, M.P. Searle and A.C. Ries, edits., Geol. Soc. Special Publication 49 : 773-786.

GORIN, G.E., RAEZ, L.G. and WELLER, M.R., 1982. Late Precambrian–Cambrian sediments of Huqf Group, Sultanate of Oman. AAPG Bulletin, 66 : 2609-2627.

GRANDVILLE, B.F. De la, 1982. Appraisal and development of a structural and stratigraphic trap oilfield with reservoirs in glacial and periglacial clastics. In: The Deliberate Search for the Subtle Trap, M.T. Halbouty, edit., AAPG Mem. 37 : 267-286.

GRANTHAM, P.J., LIJMBACH, G.W.M., POSTHUMA, J., HUGHES CLARKE, M.W., and WILLINK, R.J., 1988. Origin of crude oils in Oman. Journ. Petroleum Geol. 11 : 61-80.

GREENWOOD, W.R., 1973. The Ha'il Arch–a key to deformation of the Arabian Shield during evolution of the Red Sea Rift. Saudi Arabian Dir. Gen. Miner. Resources Bulletin, 7, 5p.

GREENWOOD, W.R., ANDERSON, R.E., FLECK, R.J., and SCHMIDT, D.C., 1980. Precambrian geologic history and plate tectonic evolution of the Arabian Shield. Saudi Arabian Dir. Gen. Miner. Resources Bulletin, 24, 35p.

GRUNAU, H.R., 1977. Generation, migration, entrapment and retention of hydrocarbons in the Middle East. Petroleum Times (June 10): 33-43.

GRUNAU, H.R., 1985. Future of hydrocarbon exploration in the Arab World. In: Source and Habitat of Petroleum in the Arab Countries, Proc. OAPEC Seminar, Oct. 1984, Kuwait, 451-499.

HAITHAM, F.M.S. and NANI, A.S.O., 1990. The Gulf of Aden Rift : hydrocarbon potential of the Arabian sector. Journ. Petroleum Geol. 13: 211-210.

HALBOUTY, M.T., MEYERHOFF, A., KING, R.E., DOTT, R.H., KLEMME, H.D. and SHABAD., T., 1970. World's giant oil and gasfields; geologic factors affecting their formation and basin classification. In: Geology of Giant Petroleum Fields, M.T. Halbouty, edit., AAPG Mem., 14 ; 502-555.

HASSAN, T.H. and AZER, S., 1985. The occurrence and origin of oil in offshore Abu Dhabi. Proc. 4th SPE Middle East Oil Techncial Conf., March 1985, Bahrain, SPE 13696, 143-155.

HAWKINS, T.R.W., HINDLE, D. and STRUGNELL, R., 1981. Outline of the stratigraphy and structural framework of southern Dhofar (Sultanate of Oman). Geol. en. Mijnb; 60 : 247-256.

HAYNES, S.J. and McQUILLAN, H., 1974. Evolution of the Zagros suture zone, southern Iran. Geol. Soc. America Bulletin, 85:739-744.

HEIMBACH, W., 1976. Jordanie. In: Leqique Stratigraphique Indternational, Vol. 111, Assie, L. Dubertret, edit., fasc. 10cl 2nd edition revised, CNRS, Paris, 150p.

HELAL, A.H., 1964. On the occurrence and stratigraphic position of Permo-Carboniferous tillites in Saudi Arabia. Geol. Rundsch., 54 : 193-207.

HEMPTON, M.R., 1987. Constraints on Arabian plate motion and extensional history of the Red Sea. Techtonics, 6 : 687-705.

HENSON, F.R.S., 1951. Observation on the geology and petroleum occurrences of the Middle East. 3rd World Petrol. Cong. Proc., The Hague, 1 : 118-140.

HEWARD, A.P., 1990. Salt removal and sedimentation in southern Oman. In: The Geology and Tectonics of the Oman Region, A.F.H. Robertson, M.P. Searle and A.C. Ries, edits, Geol. Soc. Special Publication 49 : 637-652.

HOKE, J.H. and ZIMERMAN, S., 1963. Refraction prospecting in the Rub al Khali. 4th Arab Petrol. Cong. Proc., Beirut, 26(B-2), 16p.

HUGHES CLARKE, M.W., 1988. Stratigraphy and rock unit nomenclature in the oil-producing area of Interior Oman. Journ Petroleum Geol., 11 : 5-60.

HUGHES, G.W., VAROL, O. and BEYDOUN, Z.R. (In press). New evidence for Middle Oligocene rifting of the Gulf of Aden and for Late Oligocene rifting of the southern Red Sea. Marine and Petroleum Geology.

HUSSEINI, M.I., 1988. The Arabian Infracambrian extensional system. Tectonophysics, 143 : 93-103.

HUSSEINI, M.I., 1989. Tectonic and depositional model of Late Precambrian–Cambrian Arabian and adjoining plates. AAPG Bulletin, 73 : 1117-1131.

HUTCHINSON, I., LOUDEN, K.E., WHITE, R.S. and Von HERTZEN, R.P., 1981. Heat Flow and the age of the Gulf of Oman. Earth Planet. Sci. Lett., 56: 252-262.

JAMES, G.A. and WYND, J.G., 1965. Stratigraphic nomenclature of Iranian Oil Consortium Agreement

Area. AAPG Bulletin, 49 : 2182-2245.
KADHI, A. Al-, and HANCOCK, P.L., 1980. Structure of the Durma-Nisah segment of the Central Arabian Graben System. Saudi Arabian Dir. Gen. Miner. Resour. Bulletin, 16, 14p.
KAMEN-KAYE, M., 1970. Geology and productivity of Persian Gulf synclinorium. AAPG Bulletin, 54 : 2371-2394.
KARPOFF., R., 1957. Sur l'existance du Maestrichien au Nord du Djeddah, (Arabie Seoudite). C.R. Acad. Sci. France, 245 : 1322-1324.
KAZMIN, V., RICOU, L-E., SBORTSHIKOV, I.M., 1986. Structure and evolution of the passive margin of the eastern Tethys. Tectonophysics, 123 : 153-179.
KENT, P.E. and WARMAN, H.R., 1972. An environmental review of the Worlds's richest oil-bearing region: the Middle East. 24th Internat. Geol. Congr. Proc., Montreal, 5 : 142-152.
KHIRSAN, H. Al-, and SIDDIKI, A. Al-, 1989. Hydrocarbon exploration development in Iraq. 28th Internat. Geol. Congr. Abstracts, Washington D.C., 1 : 28.
KIRKLAND, D.W. and EVANS, R., 1980, Source-rock potential of evaporitic environment. AAPG Bulletin, 65 : 181-190.
KRUCK, W. and THIELE, J., 1983. Late Palaeozoic glacial deposits in the Yemen Arab Republic. Geol. Jb. B46 : 3-29.
LABABIDI, M.M. and HAMDAN, A.M., 1985. Preliminary Lithostratigraphic Correlation Study in OAPEC Member Countries. Organisation of Arab Petroleum Exporting Countries, Kuwait, 171p.
LABOUN, A.A. Al-, 1986. Stratigraphy and hydrocarbon potential of the Paleozoic succession in both Tabuk and Widyan basins, Arabia. In:Future Petroleum Provinces of the World, M.T. Halbouty, edit., AAPG Mem. 40 : 373-397.
LA BRECQUE, J.K. and ZITTELINI, N., 1985. Continuous sea floor spreading in Red Sea: an alternative interpretation of magnetic anomaly pattern. AAPG Bulletin, 69:513-525.
LAMARE, P. 1936. Structure Geologique de l'Arabie. Libr. Politechn. Ch. Beranger, Paris-Liege, 65p.
LAUGHTON, A.S., 1966. The Gulf of Aden. Phil. Trans. Roy. Soc. London, A259:150-171
LAW, J. 1957. Reasons for Persian Gulf oil adundance. AAPG Bulletin, 41:51-69
LEES, G.M., 1951. The oilfields of the Middle East. 3rd World Petrol.Cong.Proc., The Hague, 1:94-101
LEES, G.M., 1953. The Middle East. In: The Science of Petroleum, VI(i): 73-82, Oxford University Press.
LEVELL, B.K., BRAAKMAN, J.H. and RUTTEN, K.W., 1988. Oil-bearing sediments of Gondwana glaciation in Oman. AAPG Bull; 72:775-796.
LIPPARD, S.J., SHELTON, A.W. and GASS, I.G., 1986. The Ophiolite of Northern Oman. Geol. Soc. London Mem. 11, 178p.
LOUTFI, G. and ABDEL SATTAR, M.M., 1987. Geology and hydrocarbon potential of the Triassic succession in Abu Dhabi, U.A.E. Proc. 5th. SPE Middle East Technical Oil Conf., March 1987, Bahrain, SPE 15698, 717-734.
LOUTFI, G. and BISHLAWY, S. El-, 1989. Geology and hydrocarbon potential of the deep Paleozoic sequence in Abu Dhabi. In: OAPEC/ADNOC Seminar on Deep Formations in the Arab Countries: Hydrocarbon Potential and Exploration Techniques, Abu Dhabi, Oct. 1989, Proc. Technical Papers: 85-129.
LOVELOCK, P.E.R., 1984. A review of the tectonics of the northern Middle East region. Geol. Mag., 121: 577-587.
LOVELOCK, P.E.R., POTTER, T.L., WALSWORTH-BELL, E.B. AND WEIMER, W.M., 1981. Ordovician rocks in the Oman Mountains: The Amdeh formation. Geol. en Mijnb., 60: 487-495.
MAKRIS, J. and RIHM, R., (In press). Shear controlled evolution of the Red Sea: pull apart model. Tectonophysics.
MARJEBY, A. Al- and NASH, D., 1986. A summary of the geology and oil habitat of the Eastern Flank Hydrocarbon Province of South Oman. Marine & Patrol. Geol., 3: 306-314.
MATTES, B.W. and CONWAY MORRIS, S., 1990. Carbonate/evaporite deposition in the Late Precambrian-Early Cambrian Ara formation of southern Oman. In: The Geology and Tectonics of the Oman Region, A.H.F. Robertson, M.P. Searle and A.C. Ries, edits., Geol. Soc. Special Publication 49: 617-636.
MAYCOCK, I., 1987 (Oral Account). Oil exploration and development in Marib/Al Jauf basin, Yemen Arab Republic. Proc. 5th SPE Middle East Oil Technic. Conf., March 1987, Bahrain, SPE 15685 (unpublished).
McBRIDE, J.H., BARAZANGI, M., BEST, J., Al-SAAD, D., SAWAF, T., Al-OTRI, M., and GEBRAN, A., 1990. Seismic reflection structure of intracratonic Palmyride fold-thrust belt and surrounding Arabian platform, Syria. AAPG Bulletin, 74: 238-259.
McCLURE, H.A., 1978. Early Paleozoic glaciation in Arabia. Palaeogeog. Palaeoclim. Palaeoecol., 25:315-326.
McCLURE, H.A., 1980. Permo-Carboniferous glaciation in the Arabian Peninsula. Geol. Soc. America Bulletin, 91:707-712.
METWALLI, M.H., PHILIP, G. and MOUSSLY, M.M., 1974. Petroleum-bearing formations in northeastern Syria and northern Iraq. AAPG Bulletin, 58:1781-1796.
MINA, P., RAZAGHNIA, M.T. and PARAN, Y., 1967. Geology and geophysical studies and exploratory drilling of the Persian continental shelf-Persian Gulf. 7th World Petrol. Cong. Proc., Mexico, 2:871-903.
MOUNTAIN, G.S. and PRELL, W.L., 1990. A multiphase plate tectonic history of the southeast continental margin of Oman. In: The Geology and Tectonics of the Oman Region, A.H.F. Robertson, M.P. Searle and A.C. Ries, edits., Geol. Soc. Special Publication 49:725-743.
MONTENAT, C., BLONDEAU, A., BIJOU, G., PERRAU, M., RAJU, D.S.N. and ROMAN, J., 1977. Premier apercu du Tertiaire d'Oman (Peninsule ara-

bique orientale). Bulletin Soc. Geol. France, (7e Serie), 19(6): 1285-1295

MORTON, D.M., 1959. The geology of Oman. 5th World Petrol. Cong. Proc., New York, 1:277-294.

MOSELEY, F. and ABBOTS, I.L., 1979. The ophiolite melange of Masira, Oman. Journ. Geol. Soc. London, 136: 713-724.

MOUTY, M., 1976. Sur la presence du Lias dans le massif Allaouite (Syrie), C.R. Somm. Soc. Geol. France, 73(3): 104-105.

MULDER, C.J., LEHNER, P. and ALLEN, D.C.K., 1975. Structural evolution of the Neogene salt basins in the eastern Mediteranean and the Red Sea. Geol. en Mijnb. 54: 208-221.

MURRIS, R.J., 1980. The Middle East: stratigraphic evolution and oil habitat. AAPG Bulletin, 64: 597-618.

NAKHAL, H.A. El-, 1984. Possible Late Palaeozoic glaciation in the central parts of the Yemen Arab Republic. Journ. Glaciology, 30 (104): 126-128.

NANDYAL, M., THURBER, S.S., Al-HINAI, K.M. and JAFFER, N.A., 1983. Development of heavy oil reserves in South Oman. Proc. 4th SPE Middle East Technical Oil Conf., March 1983, Bahrain, SPE 11474, 259-270.

NEBERT, K., ALSHAIBI, A.A., AWALIA, M., BOUNNY, I., NAWAB, Z.A., SHARIEF, O.H., SHERBINI, O.A. and YESLAM, A.H., 1974. Geology of the area north of Wadi Fatima, Kingdom of Saudi Arabia. Cent. Applied Geol. Bulletin 1:1-31, Jeddah.

NRA (Natural Resources Authority) Jordan, 1986. Petroleum Exploration Opportunities in Jordan, Amman, 40p.

O'CONNOR, T.E., 1990 (Oral Account). Petroleum evaluation of multinational sedimentry basins: the Red Sea/Gulf of Aden regional hydrocarbon study project. AAPG Annual Convention, June, 1990, San Francisco. Abstract, AAPG Bulletin, 74:732.

PAUL, S.K., 1990. People's Democratic Republic of Yemen: a future oil province. In: Classic Petroleum Provinces, J. Brooks, edit., Geol. Soc. Special Publication 50:329-339.

PETERSON, J.A., and WILSON, J.L., 1986. Petroleum stratigraphy of the Northeast Africa-Middle East region. In: Hydrocarbon Potential of Intense Thrust Zones, Proc., OAPEC Seminar Dec. 1986, Abu Dhabi. Vol II, 227-330.

PLATEL, J.P. and ROGER , J., 1989. Evolution geodynamique du Dhofar (Sultanat d'Oman) pendant le Cretace et le Tertiaire en relation avec l'ouverture du golfe d'Aden. Bulletin Soc. Geol. France, (8) V(2): 253-263.

PONIKAROV, V.P., KAZMIN, V.G., MIKHAILOV, I.A., RAZVALIAYEV, A.V., KRASHENINNIKOV, V.A., KOZLOV, V.V., SOULIDI-KANDRATIYEV, E.D., MIKHAILOV, K.Y., KULAKOV, V.V., FARADHZEV, V.A. and MIRZAYEV, K.M., 1967. The Geology of Syria: Explanatory Notes on the Geological Map of Syria. Scale 1:500,000 Part I, Stratigraphy, Igneous Rocks and Tectonics. Ministry of Industry, Damascus, Syrian Arab Republic, 229p.

POWERS, R.W., RAMIREZ, L.F., REDMOND, C.D. and ELBERG, E.L., 1966. Sedimentary Geology of Saudi Arabia. In: Geology of the Arabian Penisula. Geol. Surv. Professional Paper 560-D, U.S. Geol. Surv., 147p.

QIDWAI, H.A., KHALIFA, M.I. and BA-MKHALIF, K.A., 1988. Evidence of Permo-Carboniferous glaciation in the basal Murbat Sandstone Formation, southern region, Sultanate of Oman. Journ. Petroleum Geol., 11:81-88.

RICHE, P.H. and PRESTAT, B., 1979. Paleogeographie du Cretace moyen du Proche Orient et Moyen Orient et sa signigication petroliere. 10th World Petroleum Congr. Proc., Bucharest, 2:57-75.

RICHTER, H., POST, J., MAKRIS, J., GOTZ, L-G. and RIHM, R. (in press). Geophysical observations offshore Saudi Arabia: seismic, magnetic and heatflow measurements. Tectonophysics.

RICOU, L.E., 1971. Le croissant ophiolitique periarabe, une ceinture de nappes mises en place en Cretace superieur. Rev. Geog. phys. Geol. dynam. (2e serie), 13:327-349.

RIGO De RIGHI, M. and CORTESINI, A., 1969. Gravity tectonics in foothill structure belt of southeast Turkey. AAPG Bulletin, 48: 1911-1937.

ROGER, J., PLATEL, J.P., CAVELIER, C. and BOURDILLON De GRISSAC, C., 1989. Donnees nouvelles sur la stratigraphie et l'histoire geologique du Dhofar (Sultanat d'Oman). Bulletin Soc. Geol. France (8) V(2): 265-277.

ROTHSTEIN, Y. and KAFKA, A.L., 1982. Seismotectonics of the southern boundary of Anatolia, eastern Mediterranean region: subduction, collision and arc jumping. Journ. Geophys. Research, 87(B9): 7694-7706.

SAVOYAT, E., SHIFERAW, A. and BALCHA, T., 1989. Petroleum exploration in the Ethiopian Red Sea. Journ. Petroleum Geol., 12:187-204.

SCOTESE, C.R., BAMBACH, R.K., BARTON, C., Van Der VOO, R. and ZEIGLER, A.M., 1979. Paleozoic base maps. Journ. Geology 87:217-277.

SCOTESE, C.R., and McKERROW, W.S., 1990. Revised World maps and introduction. In: Palaeogeography and Biogeography, W.S. McKerrow and C.R. Scotese, edits., Geol. Soc. Memoir 12:1-21.

SCOTESE, C.R., and McKERROW, W.S., (in press). Ordovician plate tectonic reconstructions. In: Ordovician Symposium, C.R. Barnes, edit.

SENGOR, A.M.C., 1979. Mid Mesozoic closure of Permo-Triassic Tethys and its implications. Nature, 279:590-593.

SENGOR, A.M.C., 1990. A new model for the late Palaeozoic-Mesozoic tectonic evolution of Iran and implications for Oman. In: The Geology and Tectonics of the Oman Region, A.H.F. Robertson, M.P. Searle and A.C. Ries, edits., Geol. Soc. Special Publication 49: 797-831.

SENGOR, A.M.C. and YILMAZ, Y., 1981. Tethyan evolution of Turkey: a plate tectonic approach. Tectonophysics, 75: 181-241.

SENGOR, A.M.C., YILMAZ, Y. and SUNGURLU, O.,

1984. Tectonics of the Mediterranean Cimmerides: nature and evolution of the western termination of the Paleo-Tethys. In: The Geological Evolution of the Eastern Mediterranean, J.E. Dixon and A.H.F. Robertson, edits., Geol. Soc. Special Publication 17:77-112.

SENGOR, A.M.C., ALTINER, D., CIN, A., USTAOMER, T.and HSU, K.J., 1988. Origin and assembly of the Tethyside orogenic collage at the expense of Gondwana Land. In: Gondwana and Tethys, M.G. Audley-Charles and A. Hallam, edits., Geol. Soc. Special Publication 37:119-181.

SETUDEHNIA, A.O., 1972. Iran du Sud Ouest. In: Lexique Stratigraphique International, Vol III Assie,.L. Dubertret, edit., CNRS Paris, fasc. 9b:285-376.

SHACKLETON, R.M. and RIES, A.C., 1990. Tectonics of the Masirah Fault Zone and eastern Oman: In: the Geology and Tectonics of the Oman Region, A.H.F. Robertson, M.P. Searle and A.C. Ries, edits., Geol. Soc. Special Publication 49:715-724.

SHARIEF, F.A. and MOSHRIF, M.A., 1989. Chronostratigraphy and hydrocarbon prospects of the Saka-ka formation, northern Saudi Arabia. Journ. Petroleum Geol., 12:85-102.

SMITH, A.G. and WOODCOCK, N.H., 1976. Emplacement model for some Tethyan ophiolites. Geology, 4:653-656.

SMITH, A.G., HURLEY, A.M. and BRIDEN, J.C., 1981. Phanerozoic Paleocontinental World Maps. Cambridge University Press, 102p.

SOEC (Somaliland Oil Exploration Company) (Mackay, C., Williams, D., Gilbert, D.C.F.and Widmer, H.), 1954. A Geological Reconnaissance of the Sedimentary Deposits of the Protectorate of British Somaliland. Crown Agents, London, 41p.

SPIRO, B., WELTE, D.H., RULLKOTTER, J. and SHAEFFER, R.G., 1983. Asphalts, oils and bituminous rocks from the Dead Sea area–geochemical correlation study. AAPG Bulletin, 67:1163-1175.

STACEY, J.C. and AGAR, R.A., 1985. U-Pb isotopic evidence for the accretion of a continental microplate in the Zalm region of the Saudi Arabian Shield. Journ. Geol. Soc. London, 142: 1189-1203.

STERN, R.J., 1985. The Najd fault system, Saudi Arabia and Egypt: a Late Precambrian rift-related transform system? Tectonics, 4:497-511.

STOCKLIN, J., 1968. Structural history and tectonics of Iran: a review. AAPG Bulletin, 52: 1229-1258.

STOCKLIN, J., 1986. The Vendian-Lower Cambrian salt basins of Iran, Oman and Pakistan: stratigraphy, correlations, paleogeography. Sciences de la Terre, Mem. 47: 329-345, Nancy.

STOESSER, D.B. and CAMP, V.E., 1985. Pan-African microplate accretion of the Arabian Shield. Geol. Soc. America Bulletin, 96:817-826.

STONELEY, R., 1987. A review of petroleum source rocks in parts of the Middle East. In: Marine Petroleum Source Rocks, J. Brooks and A.J. Fleet, edits., Geol. Soc. Special Publication 26: 263-269.

STONELEY, R., 1990a. the Arabian continental margin in Iran during the Late Cretaceous. In: The Geology and Tectonics of the Oman Region, A.H.F. Robertson, M.P. Searle and A.C. Ries, edits., Geol. Soc. Special Publication 49:787-795.

STONELEY, R., 1990b. The Middle East Basin: a summary overview. In: Classic Petroleum Provinces, J. Brooks, edit., Geol. Soc. Special Publication 50: 293-298.

STYLES, P., and HALL, S.A., 1980. a comparison of seafloor spreading histories of the western Gulf of Aden and central Red Sea. Atti. Covegni Lincei, Accad. Naz. Lincei, 47:585-606.

SUGDEN, W., STANDRING, A.J. and CAVELIER, C. 1975. Qatar Peninsula. In: Lexique Stratigraphique International, Vol.III Asie, L. Dubertret, edit., CNRS Paris, fasc. 10b3, 120p.

SYKES, R.M. and ABU RISHEH, A.K., 1989. Exploration of deep Paleozoic and Pre-Cambrian plays in the Sultanate of Oman. In: OAPEC/ADNOC Seminar on Deep Formations in the Arab Countries: Hydrocarbon Potential and Exploration Techniques, Abu Dhabi, Oct. 1989, Proc Country Reprots and Case Studies: E71-E113.

TAKIN, M., 1972. Iranian geology and continental drift in the Middle East. Nature, 235: 147-150.

TEWFIK, N. and AYYAD, M., 1982. Petroleum exploration in the Red Sea shelf of Egypt. 6th EGPC Exploration Seminar, Cairo, 30p.

USGS (United States Geological Survey)–ARAMCO (Arabian-American Oil Company), 1963. Geologic Map of the Arabian Peninsula, Scale 1:2,000,000. U.S. Geol. Surv. Misc. Invest. Map I-270A.

VASLET, D., 1989. Late Ordovician glacial deposits in Saudi Arabia: a lithostratigraphic revision of the Early Paleozoic succession. Professional Papers No. PP-3: 13-44, Saudi Arabian Deputy Ministry for Mineral Resources, Jiddah.

WALLEY, C.D., 1983. A revision of Lower Cretaceous stratigraphy in Lebanon. Geol. Rundsch., 72:377-388.

WEBER, H., 1963. Ergebnisse erdolgeologischer Aufschussarbeiten der DEA in Nordost-Syrian; 1, Schichtfloge, fazies, tectonik in der Haute Djesireh. Erdol u Kohle, 16(6): 669-682.

WEEKS, L.G., 1958. Habitat of oil and some factors that control it. In: Habitat of Oil, AAPG Symposium, L.G. Weeks, edit., 1-16.

WELLAND, M.J.P., and MITCHELL, A.H.G., 1977. Emplacement of the Oman ophiolite: a mechanism related to subduction and collision. Geol. Soc. America Bulletin, 88: 1081-1088.

WETZEL, R., 1974. Etapes de la prospection petroliere en Syrie et au Liban. Notes et Memoires, Compagnie Francaise Petrole, 11:40-70.

WOLFART, R., 1967. Zur entwicklung der palaozoischen Tethys in vorderasien. Erdol u Kohle, 20:168-180.

WOLFART, R., 1981. Lower Palaeozoic rocks of the Middle East. In: Lower Paleozoic Rocks of the World, C.H. Holland, edit., John Wiley, 3:5-130.

WRIGHT, V.P., RIES, A.C. and MUNN, S.G., 1990. Intraplatformal basin-fill deposits from the Infracambrian Huqf Group, east central Oman. In: The Geology and Tectonics of the Oman Region, A.H.F. Robertson, M.P. Searle and A.C. Ries, edits., Geol. Soc. Special Publication 49:601-616.

ZAMEL, A., 1975. Recent exploratory development in the Middle East. 9th World Petroleum Cong. Prosc., London, 3:203-204.

ZARRUG, A.Y. and BOIS, C., 1983. Potential oil reserves in the Middle East and North Africa. 11th World Petrol. Congr. Proc., London, 2:261-275.